STATISTICAL METHODS FOR THE EARTH SCIENTIST

STATISTICAL METHODS FOR THE EARTH SCIENTIST

An Introduction

Roger Till

Department of Geology, University of Reading

A HALSTED PRESS BOOK

JOHN WILEY & SONS

New York – Toronto

First published in the United Kingdom 1974 by
The Macmillan Press Ltd

Published in the U.S.A. and
Canada by Halsted Press, a
Division of John Wiley & Sons, Inc.,
New York

Printed in Great Britain

Library of Congress Cataloging in Publication Data

Till, Roger.
Statistical methods for the earth scientist.
"A Halsted Press book."
Bibliography: p.
1. Earth sciences—Statistical methods. I. Title.
QE33.2.M3T54 550'.1'82 73–22704
ISBN 0 470–86789–2

To Libby

CONTENTS

PREFACE

This textbook is intended for use by earth science undergraduate and post-graduate students who wish to learn about the application of statistics to their subject. It is also suitable for those earth scientists who having qualified some years ago did not have a statistics course in their degree syllabus, and now feel the need to learn how to apply statistical methods in their work.

The procedures described are relevant to many branches of the earth sciences, although the majority of examples used are taken from the work of geologists. Nonetheless the examples span not only 'traditional' geological study of rocks and fossils, but wider fields including analysis of natural waters, study of the processes of weathering and sedimentation, description of landforms and measurement of engineering properties. As a result this text should be of value to the geophysicist, the physical geographer and the soil scientist as well as the geologist, since all study the same processes and materials, but from different vantage points, and all use the same statistical procedures.

The book has arisen from a course that I have been involved in teaching at Reading for the last three years. I discovered that though lengthy, expensive and advanced tomes on the use of statistics in the earth sciences existed (Miller and Kahn, 1962; Krumbein and Graybill, 1965; Griffiths, 1967; Koch and Link, 1970, 1971; Davis, 1973), no introductory texts suitable for student use were available. My aim is therefore to produce a text which can be used with a one term course of about forty hours.

Throughout the book statistical procedures are described that are needed to answer real problems that occur in the earth sciences and in all cases the procedures are worked through with a realistic data set. A 'cookery-book' approach to the use of statistics is not advocated or employed. I believe, and the literature shows, that serious errors can be made if a statistical test is applied without understanding its background and assumptions. At the same time, only simple algebra is required throughout this book. Many people have tended to shy away from statistics because of the unfamiliar mathematical shorthand it employs. I have attempted to show that if you can understand the form and nature of a population, the concepts of probability, and the use of sample estimates of population parameters, then all statistical tests become many variations on the same theme.

All the calculations in this book can easily be performed on a small mechanical or electronic calculator. Most students would nowadays progress to using a computer, either through an on-line terminal or a 'cafeteria-type' of service. It is important to start with some form of manual calculation since you learn more about a statistical procedure by working through its computation.

I have not provided an appendix of statistical tables because I think that just as you should have a set of log tables so you should have a pocket set of statistical tables. The comprehensive books of tables by Fisher and Yates (1948) and Pearson and Hartley (1969, generally referred to as the *Biometrika* tables) will be available in all your libraries. In the text I have produced graphs of functions which that are not usually found in the pocket sets of statistical tables.

The first three chapters of the book deal with the basic theory of statistics—measurement, probability and the nature of populations. Building on these, the next four chapters deal with the actual application of statistical tests (to data sets we are likely to collect), sampling, testing samples, t-tests, χ^2 tests, correlation and regression, and F-tests. The last of these chapters is devoted to non-parametric statistical methods. The nature of many observations in the earth sciences makes these last methods important although the whole field of non-parametric testing has been neglected in previous texts. The most commonly used non-parametric test for nominal and ordinal data are described, followed by a consideration of correlation methods. The final chapter gives a short, non-mathematical guide to the more advanced statistical procedures. In the relevant chapters a few references are given of examples of good statistical usage in the earth science literature. I hope that readers will go on from this book to read these papers and work through the data they contain and that this book will provide a basis for moving on to the more advanced books I referred to earlier, and also for dipping into some non-earth science statistical texts. I have found those by Fryer (1966), Li (1964), and Yule and Kendall (1953) readable and very useful.

Many friends and colleagues have helped me with this book. John Allen, Roger Anderton, Ken Bailey, Dave Hopkins, Tom Huntingdon, Andrew Parker, Tom Prudence, Robert Sawyer and John Thomas here at Reading kindly provided me with sets of data. So did Paul Bridges (at Derby College of Art and Technology), Charles Curtis (at Sheffield University), Colin Dunn (at Department of Mineral Resources, Regina, Canada), Mike Leeder (at Leeds University) and Pat Wilson (at Kings Langley School, Hertfordshire). These people are all acknowledged at the appropriate places in the text. Perce Allen kindly allowed me to base my derivation of the binomial distribution on one of his former lectures. I have had many useful discussions with John Allen and John Thomas, both of whom have always helped and encouraged me. Clive McCann and Dave Hopkins taught with me on our undergraduate course and we have all learnt a lot from this collaboration and attempt at team-teaching, as I hope did the students!

I am particularly grateful to Clive McCann for his friendship and encouragement. He also helped by reading and offering useful advice on several sections of the book. Don Woodrow (at Hobart and William Smith Colleges, New York) kindly read and commented on an early draft of part of the book. My interest in statistics was fostered by contact with Charles Curtis and Alan Spears when I was a research student at Sheffield University. This early enthusiasm was vital to me.

The facilities and environment in which to write this book have been made readily available to me by Perce Allen here at Reading. I am greatly indebted to him. My thanks go to Linda Barber, Cynthia Daulby, Christine Jones and Ann Fuller, who all dealt uncomplainingly with my handwriting to type the manuscript. My wife has helped and inspired me greatly in this project.

Reading ROGER TILL

1 STATISTICS AND MEASUREMENT IN THE EARTH SCIENCES

The word statistics was first used in 1770, but with a rather different meaning from that used today. One chapter of Hooper's *The Elements of Universal Erudition* published in 1770 is entitled 'Statistics' and deals with 'the science that teaches us what is the political arrangement of all the modern States of the known world' (Yule and Kendall, 1953). In the early decades of the nineteenth century the change to 'statistics' representing the characters of a State by numerical methods was taking place. Only by the end of the century were 'statistics' the summary figures used to describe and compare the properties of a set of observations. At about this time the theoretical basis of the science of statistics was being laid, and today we find the ideas of statistics on a firm basis and applied to the collection, summary and analysis of all types of data.

The degree of quantification employed has varied markedly in the different branches of the earth sciences. Geophysics has always been a highly numerate, mathematical subject involving precise measurements of gravity, magnetic fields, heat flow, spreading rates and other field and laboratory observations, using complex procedures, such as Fourier analysis.

The use of statistical procedures in physical geography, soil science and geology have developed together. This is because all three disciplines are closely interwoven and all tend to be dealing with the same type of information. One could regard physical geography and soil science as dealing with some aspects of the geology of the recent! For a long time these branches of the earth sciences were considered to be merely descriptive sciences, since they only deal with scattered surface observations of processes and objects which have taken millions of years to produce. Consequently any quantification was felt to be unreliable or pointless. It is only with the rise of geophysics and the application of other scientific disciplines, particularly physics and chemistry, that they have become more numerate and quantitative subjects. As explained in the preface, most of the examples used in this text come from the work of geologists, but many are in the domain of the physical geographer and the soil scientist.

Some of the earliest applications of statistical procedures to geology took place in the 1930s when sediments started to be analysed in a quantitative manner (Krumbein and Pettijohn, 1938). During the next thirty years statistical procedures were applied to all aspects of the subject. In palaeontology Kermack, Miller and Simpson; in geochemistry Ahrens, Chayes and Vistelius; in sedimentology Allen, Griffiths, Krumbein and Potter; in geomorphology Chorley, Horton, Scheidegger and Strahler have developed statistical procedures, to mention a few. This quantitative approach did not gain great support, partly because of the background of most geologists, partly because of the new jargon required to understand the procedures and partly because some of the proposed statistical techniques were inapplicable to the earth sciences.

Since the early 1960s the floodgates of quantification have been opened. This deluge has been aided and abetted by the availability of fast electronic digital computers. For a while this produced an attitude of mind in which all data was analysed statistically on the computer. No consideration was given to the origin of the data, or the requirements or purposes of the tests. The situation is settling down now and most earth scientists receive some training in statistics in their undergraduate or postgraduate courses.

A number of advanced books applying statistics to geology are now available (Miller and Kahn, 1962; Krumbein and Graybill, 1965; Koch and Link, 1970, 1971; Harbaugh and Bonham-Carter, 1970; and Davis, 1973) and many more research papers employ statistical methods. Doornkamp and King (1971) describe the use of statistical methods in geomorphology, though only in a 'cookery-book' approach. The geographers' journals *Area* and *Zeitschrift für Geomorphologie* often have statistical papers in them, as do *Soil Science* and *Journal of Soil Science*, with which soil scientists will be familiar. Perhaps 'quantitative methods in geology' has come of age, especially now that we have a Society—*The International Association for Mathematical Geology* and a journal *Mathematical Geology*. However there still are problems in applying statistical procedures in geology, because of the strange nature of much of geological data with its many different sources. This makes sampling procedures difficult and requires the use of some tests whose theoretical background is less well understood.

1.1 The measurement process

Any method of measurement is applied to obtain a numerical value for some property possessed by an object or a group of objects. We can make a measurement at all sorts of magnitudes from the radius of the earth to bond length in a chemical compound. Our measurements may be made with great precision, such as obtaining the salinity of a standard sea-water sample, or may only be semiquantitative, such as a geochemical orientation survey.

In any study the quality of the resulting conclusions depends on the quality of the numerical data being used. Hence it depends on the quality of the original measurement made and the care of the measurer. Subsequent analysis and interpretation of the data can only be as good as the original measurement. For this reason we must take great care in the design of sampling procedures and in the collection of our data.

Precision and accuracy

These are two very important properties of any measurement process and are treated more fully in section 4.6. Precision refers to the repeatability of a measurement. The more precise the method the more closely clustered about the average value will be the set of measurements. There is often a balance to be sought between the cost of making a measurement and its precision. Similarly the purpose of the measurement also governs the precision required.

A measurement can be very precise but also very inaccurate. The accuracy is an estimate of how close to the true value is our value, that is how unbiased is our measurement. Often we have a problem of not knowing what the true value should be, but on occasions reference can be made to artifical standards. We could measure the length of a 500 m slope section with great precision, to the nearest millimetre, yet if our tape measure is not correctly marked out, we could have a precise but inaccurate measurement.

Scales of measurement

Whenever we make a measurement we use a set of rules to define the result. These include the units and the procedure employed. Measurements grade from semi-quantitative (for example, a steep angle of repose) to highly quantitative (angle of repose of 14.35°). These very different levels of numeracy have different properties and allow us to use different types of statistical test. We can define four scales of measurement, with cumulative properties—each has the attributes of the previous scale together with some additional ones.

1. *The nominal (or classificatory) scale* is the weakest level of measurement. Numbers, symbols or names are used to classify objects using a pigeon-holding procedure. Classification of soils, rocks, minerals or fossils are examples of this type of measurement. The symbols used to represent the different states are quite arbitrary. For example, in describing the colour of sandstones

white	yellow	grey	red	could be coded
1	2	3	4	or
45	3	20	1	or
*	:	%	;	etc.

Simple arithmetic can be done and statistics calculated from these symbols. The number of red sandstones could be counted and compared to the number of yellow sandstones in a Triassic section. These methods are described in chapter 7.

If we are considering igneous rock classification, then we could take these pigeon-holes—granite, adamellite, granodiorite, syenite, diorite, gabbro and peridotite. This sort of classification is used in many parts of the earth sciences and it presents a paradox. Though the final classes are only nominal, more rigorous measurements (the quartz content of the rock or the shape of a fossil) may have been made to define the class. It is important to use the correct level of measurement for the question being posed. In many cases the recognition of granite and adamellite will solve the geological problem. Merely having a named fossil present in a rock may precisely define its age. A criticism raised at proponents of quantification in the earch sciences is that they give numbers to things just for the sake of it. This must be avoided.

2. *The ordinal (or ranking) scale* is used when objects can be recognised as different and also be put into some sequence or relationship to each other. Though the objects can be ranked or classed, the length of the steps between classes need not be constant. A good example is the Moh's hardness scale for minerals. Classes are numbered 1 to 10, but the difference between the hardness of diamond (10 on the scale) and corundum (9) is far greater than between corundum and talc (1). Provided the true rank order is preserved any number can be used to represent the class. For example

gravel	sand	silt	clay	could be
1	2	3	4	or
1	52	279	856	or
4	3	2	1	etc.

3. *The interval scale* is only established when there is equality of length of steps between classes. This makes the ratio of any two intervals independent of the units of measurement and the zero point, which are both arbitrary. Temperature is measured on this sort of scale. Though centigrade and Fahrenheit have different zeros, the ratio of any two temperature differences (that is, intervals) will be the same on both scales

°C	0	20	50	100
°F	32	68	122	212

$$\text{in } ^\circ\text{C} \quad \frac{100 - 50}{20 - 0} = 2\tfrac{1}{2}$$

$$\text{in } ^\circ\text{F} \quad \frac{212 - 122}{68 - 32} = 2\tfrac{1}{2}$$

4. The ratio scale has a true zero in addition to the properties of the interval scale, making the ratio of any two scale points independent of the units of measurement. Mass, length, stream velocity, dip and strike are all on this scale. The ratio of the measurements of any length in centimetres to its measurement in inches is always constant (and equal to 2.54); the ratio of two actual measures of a temperature, one in °C the other in °F is not.

The ratio scale is the most versatile and powerful measurement scale, because it has the maximum information and allows the most rigorous statistical tests to be used. People aspire to the highest scale of measurement possible from their data. It is possible to expend much time and effort in this pursuit unnecessarily, since very useful results can be obtained even with nominal data. This is fortunate since the best data available in parts of the earth sciences is only at this weakest measurement level.

We refer to two different sorts of measurements, attributes and variables. Attributes are recorded on the nominal or ordinal scale and have discrete values, such as 'present' or 'absent' or 'five'. Variables belong to the interval or ratio scales and have a continuous scale of values (see Yule and Kendall, 1953, p. 1–18).

Parametric and non-parametric statistics

These two main types of statistics are designed to deal with data on the different measurement scales. Parametric statistics, which specify conditions about the parameters of the population from which the sample was drawn, occupy the larger part of this book. These are the most powerful statistics (but they make the most assumptions about the distribution of measurements) and are applied to interval and ratio scale measurements—the variables.

Non-parametric statistics, mostly applied to nominal and ordinal measurements—the attributes, are considered in chapter 7.

1.2 The purpose of measurements in geology

We can think of many different types of measurements made in geology. The dip of a bed of rock, the thickness of a soil profile, the angle of extinction of a thin-section of pyroxene, the length of a brachiopod, the diameter of a spore, the potash content of a granite, the plastic limit of a clay, the temperature of a lava flow are only a few examples. These measurements are all made on some geological specimen to help us to define a geological structure, to name a fossil or to understand how a rock was formed.

Following Krumbein and Graybill (1965, chapters 1–3) we can recognise several sorts of geological data by its source. All the measurements are made in the field or in the laboratory and most of the data in both places is observational. A little is experimental, such as finding the current speed required in a flume to produce a particular bed form in a sand, or determining the pressure/temperature conditions under which a given rock will melt. These experiments are performed to help with field observations. If we observe the same bed form in the rock record it is hoped that we can deduce the current regime in which it formed.

Observational data is either qualitative (a green sandstone) or numerical (dipping at 17° in a direction 40° E of N). The most we can do with qualitative data is convert it to some sort of nominal scale (number of green sandstones, or number of rocks containing the mineral glauconite). All observational data suffers from certain drawbacks. Measurements can only be collected rather irregularly—where rocks are exposed, where currents are active or where bore-holes have been sunk. In addition many variables may be operating and interacting in a natural environment, so it is difficult to isolate the effect responsible for our measurement. These problems reinforce how careful we must be about our sampling procedures.

Having obtained a set of good measurements these tend to be used in two ways. Firstly, we can make geological interpretations directly from them. For example, a dip reading helps us trace out a field boundary, or the modal quartz content helps in naming an igneous rock. We may also record a real variation in a property and use it directly to describe the geological environment (for example, change in cleavage direction in a slate towards an intrusion). This direct geological interpretation is the most usual use of numerical or quantitative data in geology.

The second use of our measurements, which is the concern of this book, is to calculate statistics. These may be either descriptive statistics or those used for making statistical inferences. Descriptive statistics are an adjunct to making geological interpretations. Instead of using the raw observations we process them and produce numerical summaries. We calculate average grain size of a sediment, mean chromium content of an ore and an estimate of its likely variability in that ore. Then we might take our small group of measurements, our sample, and use them to typify a larger geological phenomenon.

We make statistical inferences to try and answer questions of the sort: Is the plasticity of this clay different from that of other Tertiary clays? Is the average rubidium content of this batholith the same as that of the other Caledonian batholiths in Scotland? This type of inference compares one property for several groups of specimens. Often we measure several properties (variables) on each specimen. Then we may be interested in inferring the relationship between porosity and grain size in a sandstone, or between mineralogy and melting point of a rock.

In the following chapters both descriptive and inferential statistics will be used on numerical measurements taken from all types of geological observations. It is hoped that readers will also apply the methods to data sets of their own.

2 PROBABILITY

Central to the study of statistics is the concept of probability, which may be difficult for students of the physical sciences to grasp, because of their deterministic science background. In this chapter we look at the ideas of discrete probability and develop these into continuous probability distributions in the following chapter.

The probability, p, of an event occurring will lie between zero and one, by definition. If $p = 0$, then that event will never occur, but if $p = 1$ that event will always occur in a trial. A trial is a procedure, like tossing a coin, during which the event could occur. If $p = 0.7$ then an event is likely to occur, but if $p = 0.1$ an event is not very likely to occur. We can use qualitative expressions such as likely, or very likely, but defining a value for p tell us exactly how probable any event is. For if $p = 0.7$ we would expect that event to occur seven out of ten times, on average, or seventy out of one hundred times, on average. Likewise we would expect that event not to occur, on average, for 30 per cent of trials. Note the repeated use of 'on average', since one does not expect the event to occur seven out of every ten times. Indeed if it did occur so regularly we might be suspicious.

To follow this idea further we can use some simple non-geological examples (this is the only time we shall need to). They relate to games of chance, familiar to all. It is because of man's interest in games of chance that the science of probability was so well developed by the beginning of the twentieth century.

Let us consider the process of tossing a coin (in common with all statistics texts!). If we have a fair, unbiased coin the chances of it landing heads or tails are equal. We say the probability of a single toss resulting in heads (the event) is $p = 0.5$ (or a 50 per cent probability, when we convert the zero-to-one scale into percentages). Every time the coin is tossed the probability of a head = 0.5. Each event is entirely independent of the previous one. Over a large number of tosses we would expect p to approach 0.5 quite closely, if the coin is unbiased. Table 2.1 shows the result of tossing a coin 100 times. The actual proportion of heads recorded varies quite markedly in each group of ten tosses, but over the hundred tosses approaches 0.5 more closely.

Here we have shown the simple, although difficult to accept, properties of

Table 2.1 One hundred tosses of a fair coin, recorded in groups of ten tosses

Heads	Tails	Proportion of heads
4	6	0.4
4	6	0.4
7	3	0.7
5	5	0.5
5	5	0.5
6	4	0.6
5	5	0.5
6	4	0.6
7	3	0.7
4	6	0.4
53	47	0.53

probability. By defining a probability, $p = 0.5$, we are saying that over an infinite number of tosses of a fair coin we should get an equal number of heads and tails. However, the outcome of any single trial is uncertain, and is independent of any previous or any succeeding events. Each event is equally probable to be a head or a tail, because each event is a separate entity whose outcome results only from that flick of the coin.

Our definition of probability means that we can never really determine a p value by an experiment, since we cannot perform an infinitely long experiment. In practice we would use as large a number of trials as is possible.

Having explained (a) the idea of the uncertainty of any given trial, and (b) the known probability of an infinite number of trials, we can now proceed to state a few basic rules of probability. Much fuller descriptions of probability can be found in many statistical texts devoted to the subject (for example, Von Mises, 1939, 1964; Feller, 1968). Both Fisher (1956) and Moroney (1970) give readable discussions of probability.

2.1 The probability rules

To introduce a little mathematical shorthand, the probability p, of an event E, occurring is written classically as

$$p = \Pr[E] = \frac{\text{total number of occurrences of event}}{\text{total number of trials}}$$

The probability of the event E, not occurring is q, where

$$q = \Pr[\text{not } E] = 1 - p$$

$$p + q = 1$$

Our empirical definition of p can be replaced by an alternative

$$p = \frac{\text{number of ways event } E \text{ could happen}}{\text{total number of possible outcomes of trial}}$$

For example the probability of throwing a six with one throw of a fair dice is $p = \frac{1}{6}$, since a 'six' is one event out of six possible outcomes.

Probability of compound events

E_1 and E_2 are two events and the event that both E_1 and E_2 occur is written $\Pr[E_1 E_2]$ and is calculated

$$\Pr[E_1E_2] = \Pr[E_1] \times \Pr[E_2|E_1] \tag{2.1}$$

where $\Pr[E_2|E_1]$ is the probability of 'E_2 occurring given that E_1 has taken place'. If E_1 and E_2 are independent, that is the occurrence of one does not affect the probability of occurrence of the other

$$\Pr[E_2|E_1] = \Pr[E_2]$$

$$\Pr[E_1E_2] = \Pr[E_1] \times \Pr[E_2] \tag{2.2}$$

which is a rather longwinded way of saying that the probability of two independent events occurring is the product of their individual probabilities.

For example, what is the probability of two successive tosses of a coin being heads?

$$\Pr[E_1] = 0.5, \ \Pr[E_2] = 0.5$$

and since the individual tosses are quite independent

$$\Pr[E_1E_2] = 0.5 \times 0.5 = 0.25 \text{ or } \tfrac{1}{4}.$$

We can see this result empirically. If we toss a coin twice we can have four pairs of results; head–head, head–tail, tail–head or tail–tail. Our required occurrence is one out of four possibilities. This procedure is easily extended

$$\Pr[E_1E_2E_3] = \Pr[E_1] \Pr[E_2] \Pr[E_3] \tag{2.3}$$

If E_1 and E_2 are not independent then we must use equation 2.1. For example, what is the probability of picking one blue and one green marble out of a bag containing three blue and five green marbles?

$$\Pr[E_1] = \Pr[\text{drawing a blue marble}] = \tfrac{3}{8}$$

but

$$\Pr[E_2|E_1] = \Pr[\text{drawing a green marble after a blue one}] = \tfrac{5}{7}$$

$$\Pr[E_1E_2] = \tfrac{3}{8} \times \tfrac{5}{7} = \tfrac{15}{56}$$

because the first event affects the probability of the second.

A whole set of such probability rules can be defined, but those above are the only ones we shall need. Spiegel (1961, chapter 6) describes and exemplifies a number of these rules.

Permutations and combinations

Anyone who has tried the football pools will be familiar with these terms, though perhaps unaware how to calculate them. We start by saying that if one event can occur in any one of d_1 ways, followed by a second event in d_2 ways, the number of ways in which both events occur is $d_1 d_2$. For example, if we have five large marbles and three small marbles all of different colours, there are 5×3 ways of picking a large and a small marble.

A permutation is an arrangement of r out of n objects, when the order of arrangement is important. If we choose a sample of two objects out of the total n objects, the choice of A followed by B may have a different effect from B followed by A. The total ways of making this selection would be the permutation of n objects taken two at a time. If we are only interested in the objects in our sample, that is the presence of A and B, and not their order of choice, we refer to this as a combination. That is, a selection of two objects out of the total n.

The total number of permutations of n objects taken r at a time is written

$$_n\mathrm{P}_r$$

where

$$_n\mathrm{P}_r = \frac{n!}{(n-r)!} = n(n-1)(n-2)\ldots(n-r+1) \tag{2.4}$$

and $n!$ is called the 'factorial of n'

$$n! = n \times (n-1) \times (n-2) \times (n-3)\ldots 3 \times 2 \times 1$$

and $0! = 1$ by definition .

The total number of combinations of n objects taken r at a time is written

$$_n\mathrm{C}_r \text{ or } \binom{n}{r}$$

where

$$_n\mathrm{C}_r = \frac{n!}{r!(n-r)!} = \frac{_n\mathrm{P}_r}{r!} \tag{2.5}$$

As an example how many ways are there of choosing a cricket team ($r = 11$) out of a first year geology class of thirty ($n = 30$)?

Since order of choice is not important we need $_n\mathrm{C}_r$

$$_n\mathrm{C}_r = \frac{30!}{11!(30-11)!} = \frac{30!}{11!19!}$$

$$= 54\,627\,300$$

What is the number of permutations of the musical notes ($n = 5$) A, B, C, D and E taken three ($r = 3$) at a time?

Here we are interested in order because BAD is a different tune from DAB, BDA, DBA, ABD or ADB. So

$$_{n}P_{r} = \frac{5!}{(5-3)!} = \frac{5!}{2!} = 60$$

If we were not interested in the order of notes

$$_{n}C_{r} = \frac{5!}{3!(5-3)!} = \frac{5!}{3!2!} = 10$$

These two results are easy to check by writing out the possible combinations and then realising that each group of three letters can be arranged internally in six different ways.

It is true that these examples are rather abstract and may seem trivial to the earth sciences. However, we shall need to use these probability rules in defining the binomial distribution in the following chapter. In addition we must understand the concept of probability—that the outcome of a given trial, or measurement is uncertain, but its frequency of occurrence in an infinite number of trials can be defined precisely. The following analysis of a sedimentary sequence exemplifies this concept.

2.2 Transition probability matrices

Figure 2.1 shows part of a section in the fluviatile Old Red Sandstone from South Wales. It consists of a sequence of fining-upwards cyclothems, but how can we define these quantitatively? We can recognise certain facies which tend to follow each other, but there is also a 'random element' in the sequences, making the outcome of a given event, the change from one facies to another, uncertain. We could form an ideal cyclothem, but many of the cycles do not contain all the possible transitions. What weight should we give to each of the five sedimentary facies represented?

A good way of defining these cycles is to construct an upward-transition probability matrix. Starting at the base of our section we make a tally of each change from one facies to the next. So there is one transition from A to B_3, one from B_3 to D, one from D to B_2, etc. Table 2.2 shows the totals we obtain for the ten cycles. For example, there are three transitions from facies C to facies D and seven from B_3 to D. By dividing the number of transitions in each cell by the total number of transitions for that facies type (that is, the row totals) we produce a matrix of relative frequencies. This tells us how frequently a given facies is followed by another particular facies. If we are satisfied that we have a representative section with sufficient transitions, we can regard this matrix of relative frequencies as a matrix of probabilities (table 2.3)—an upward-transition probability matrix.

Table 2.3 now tells us that if we are in facies D we will most probably pass

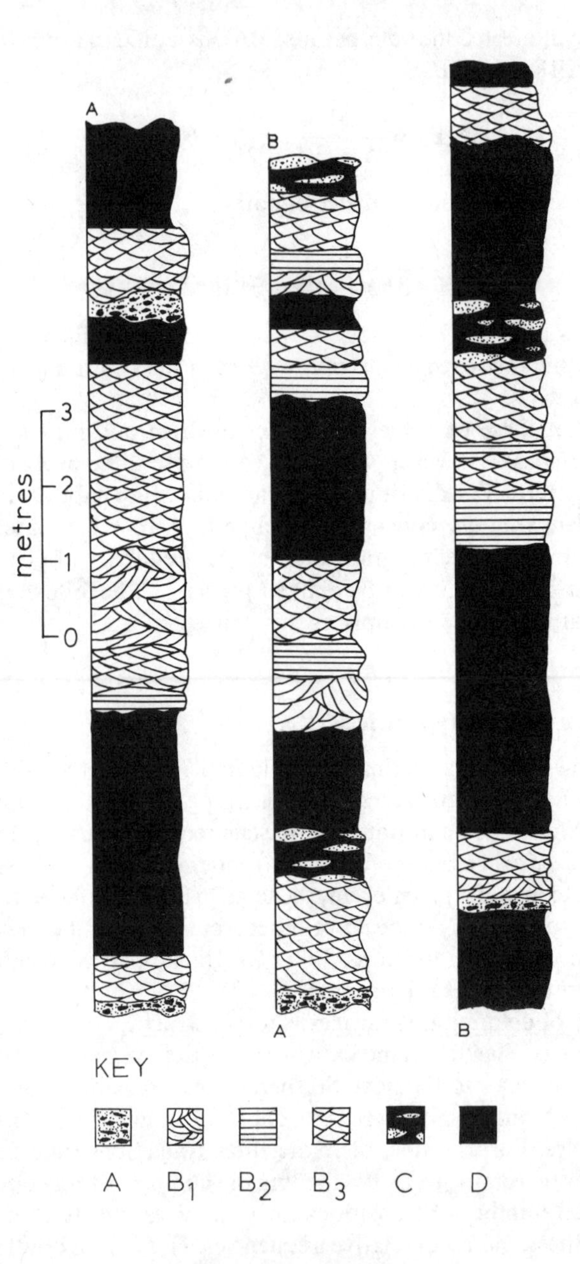

Figure 2.1 Part of a section in the Lower Devonian Sandstone-and-Marl Group at Freshwater West in Pembrokeshire, South Wales. In the key: A—*Conglomeratic Facies;* B_1—*Cross-bedded Sandstone Facies;* B_2—*Flat-bedded Sandstone Facies;* B_3—*Cross-laminated Sandstone Facies;* C—*Alternating-beds Facies;* D—*Siltstone Facies. Data provided by J. R. L. Allen.*

Table 2.2 Number of upward transitions from one sediment facies to another using the data from figure 2.1.

	A	B_1	B_2	B_3	C	D	Totals
A	—	1	0	3	0	0	4
B_1	0	—	2	2	0	0	4
B_2	0	0	—	6	0	0	6
B_3	0	1	1	—	3	7	12
C	0	0	0	0	—	3	3
D	3	2	3	1	0	—	9
Totals	3	4	6	12	3	10	38

Table 2.3 Upward-transition probability matrix for the data from figure 2.1

	A	B_1	B_2	B_3	C	D
A	—	0.25	0	0.75	0	0
B_1	0	—	0.5	0.5	0	0
B_2	0	0	—	1.0	0	0
B_3	0	0.08	0.08	—	0.25	0.59
C	0	0	0	0	—	1.0
D	0.33	0.23	0.33	0.11	0	—

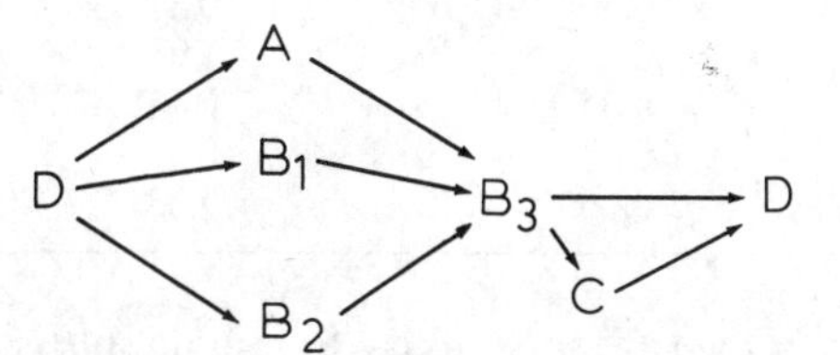

Figure 2.2 A tree diagram, summarising the most probable sequence of sediment types, constructed from table 2.3.

to facies A or B_2 ($p = 0.33$), but it we are in facies C we will certainly pass to facies D ($p = 1.0$). In this way figure 2.2 is built up to record the most probable sequences of facies to be found in the set of cyclothems. Thus if we start at D (the finest-grained member of the sequences), our most probable series of transitions is

$$D \rightarrow A \rightarrow B_3 \rightarrow D$$

or

$$D \rightarrow A \rightarrow B_3 \rightarrow C \rightarrow D$$

or

$$D \rightarrow B_2 \rightarrow B_3 \rightarrow D$$

etc.

Although the sequence of events has a random element, we have quantified the most probable forms of our cyclothem and can now make comparisons with cyclothems from other areas of fluviatile sedimentation. The process controlling the formation of our cyclothem is called a stochastic process—which is defined as a sequence of events defined by probabilities, but each event has a random element in it.

In our original tally-table (table 2.2) we did not record a transition from a facies to itself. This could occur if we had been able, or wanted, to define clear bedding planes, erosive contacts or other clear breaks within a facies. We were interested in the sequence of sedimentary environments and the sequence of depositional changes they represent and consequently only recorded transitions from one facies to another. The probability matrix can also be built-up by recording transitions on a fixed interval. In this way transitions from a facies to itself could occur. Table 2.4 shows a probability matrix generated from figure 2.1 using a fixed interval of 0.5 m. We find that

Table 2.4 Upward-transition probability matrix constructed from figure 2.1 using a constant interval of 0.5 m.

	A	B_1	B_2	B_3	C	D
A	0	0	0	1.0	0	0
B_1	0	0.5	0.2	0.3	0	0
B_2	0	0	0	1.0	0	0
B_3	0	0.04	0.04	0.48	0.13	0.31
C	0	0	0	0	0.25	0.75
D	0.1	0.06	0.1	0.03	0	0.71

if we begin in facies D we are most likely to remain in that facies, because D has the greatest thickness in the section.

The fixed interval approach is not of use to us in our example, but there could be studies where it was required (for example, recording the height of a river every month). The results of such an analysis depend on the intervals used—too large an interval misses many quick changes and too small an interval produces too many transitions from a state to itself. Furthermore, in a sedimentary sequence we cannot equate thickness with time because rates of deposition vary markedly for different sediment types and bedding planes can represent longer time periods than the actual sediment (for example, see Ager, 1973).

Markov processes

In defining possible cyclothems in our example, we are implying that the occurrence of a given facies depends on the previous facies (or the previous few facies). The sedimentary process has a memory. Such processes are called

Markov processes, so named after the Russian mathematician who studied them at the beginning of the century.

A Markov process is defined as a natural 'process which has a random element, but also exhibits an effect, in which previous events influence, but do not rigidly control, subsequent events' (Harbaugh and Bonham-Carter, 1970, p. 98). A Markov chain is a Markov process in which the probability of transition from one discrete state to the next state in the chain depends on the previous state.

This chain is said to be stationary if the probabilities associated with the transitions are constant through time. If the next step in a chain depends only on the previous state it is called a first-order Markov chain. If the next step also depends on states further back in time it is a higher-order chain.

Many natural processes are Markovian. A river tends to remain at its present state; the weather tomorrow will most probably be like the weather today. In geology, the greatest use of Markov chains has been in analysing stratigraphic sequences, starting with the procedures previously explained. A statistical test to establish the Markov property is described in section 4.5.

2.3 Applications

Further details about Markov processes can be found in Krumbein (1967), Harbaugh and Merriam (1968) and Harbaugh and Bonham-Carter (1970).

Allen (1970) uses upward-transition probability matrices to analyse fluvial cyclothems from the Old Red Sandstone. Selley (1969) uses similar matrices as a starting point for describing a variety of sedimentary sequences. He compares the observed transitions with those expected under a random arrangement. The discussion following Selley's paper highlights some of the problems seen by sedimentologists in this type of mathematical treatment.

Read (1969) applies Markov methods in his consideration of coal-bearing Namurian sediments from central Scotland. This paper stems from earlier detailed considerations of the same sediments (Read and Dean, 1967, 1968) in which a variety of statistical procedures are applied to describe these coal-bearing cyclothems. Miall (1973) applies Markov chains to Devonian sediments from Arctic Canada.

3 SOME DISTRIBUTIONS AND THEIR PROPERTIES

Let us suppose we want to determine the porosity of a Permian sandstone encountered in a large number of wells located in the North Sea. Having measured this property, using some appropriate technique, we might locate the value obtained on a map to see how the porosity of the sandstone varies over the whole basin. In addition, we may want to characterise the whole sandstone body by some sort of average, together with a parameter to define the spread of porosity values. This variation in the porosity is a real property. It is one of the characteristics of the sandstone body and might for example, relate to variations in grain size and cementation of the rock.

The porosity of an individual sandstone sample, that is a single observation, tells us little about the whole sandstone body. It could be near the average of all our values, or it could be the most extreme value recorded. We can only say how typical it is of the whole body by relating it to what we know about the distribution of values we have obtained. If we have sufficient information about our distribution, we can determine the likelihood, or probability, of a given porosity value occurring.

We have immediately come to the basic difference between a deterministic science and a probabilistic one—statistics. In chemistry we might measure the concentration of a solution. There is only one value of this for a given beaker of solution, and it completely defines that property of the solution. The porosity of our sandstone has a real, inherent variation. Any value that we use to describe that rock body can only convey part of the information. Furthermore it is only one of a large number of possible and correct values. To fully convey the porosity of this sandstone body requires the use of several descriptors, or statistics, of the set of observations.

On virtually every occasion that we make a set of measurements, we will get a distribution of values, whether it be the lengths of brachiopods in a collection, grain size of a sediment, oxygen content of a lake's water or residual angle of internal friction of a set of shale samples.

We obtain a distribution by making a single observation on each of a large number of samples. However, if we took one of our sandstone samples and repeatedly measured its porosity, we would also get a distribution of values. The method will be subject to experimental error, in just the way that we introduce errors when we read a thermometer or make up a volumetric flask. If the method is precise, all the values will be very close to each other. If not, the values may be widely spread. Such a distribution of replicate analyses is often produced to assess the precision or accuracy (see section 4.6) of a method of analysis.

Even to begin to think like real statisticians we must remember this inherent variation in observations. Nothing is certain, but we will be able to say just how uncertain things are.

3.1 Representing and describing distributions

Graphing a distribution

When we have collected a large number of observations to define a distribution we need some compact way of conveying this in our report or paper. A huge table of values is cumbersome and incomprehensible and instead some sort of frequency diagram, as shown in figure 3.1, is usually drawn.

In these histograms the observations are split into classes or intervals (grain size from $\frac{1}{2} - 1\phi$, silica content from 72.8 to 73.0 per cent). These classes are usually of equal size, though they do vary in the earthquake example (figure 3.1D) to make the diagram of a reasonable length. Observations are either collected in the classes used (weight of sediment caught on a given sieve) or grouped into them (silica content between 72.8 and 73.0 per cent) to produce the diagram. The abscissa records absolute frequency (actual weight of sediment in that class or number of occurrences of earthquakes in a given depth range), or relative frequency (proportion of all the observations that fall in the given class).

These frequency distributions are commonly represented in a variety of ways, as shown in figure 3.2 for the sediment grain size histogram (data in table 3.1). In all cases it is necessary to indicate the number of observations from which the diagram is constructed. This is especially important in a relative frequency diagram which could be produced from ten or ten thousand observations.

If we measured more and more grains of sediment we could make our size classes narrower and narrower. Eventually the classes would become so narrow that a continuous distribution would result, because an infinite number of sediment grains would contain specimens of all conceivable sizes. Figure 3.2C shows that we often pretend to have this continuous distribution by interpolating a smooth curve through the frequencies at the

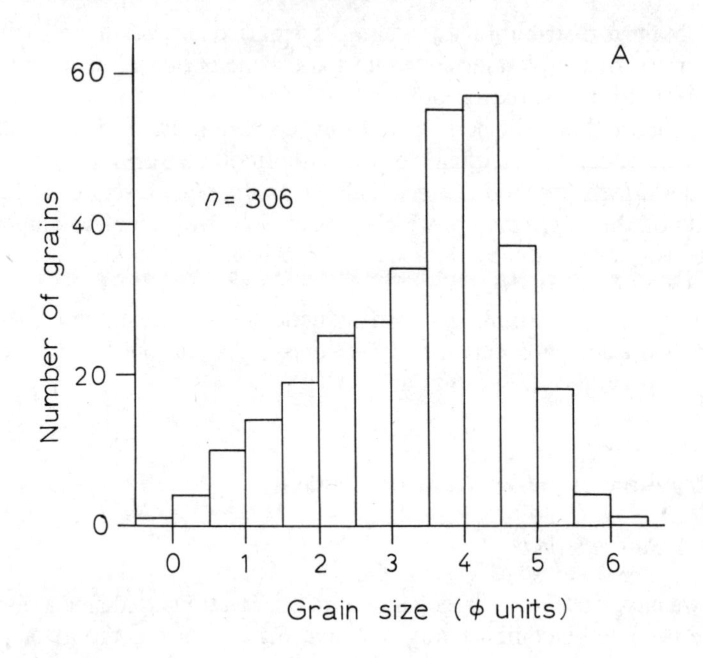
A
Number of grains
60
40
20
0
n = 306
0
1
2
3
4
5
6
Grain size (ϕ units)

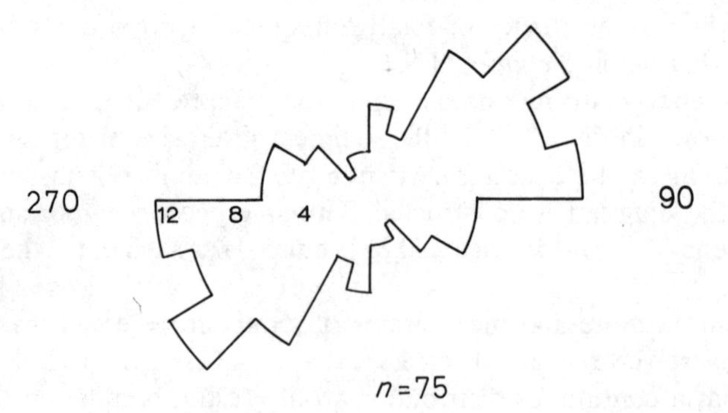
0
B
270
90
12
8
4
n=75
180

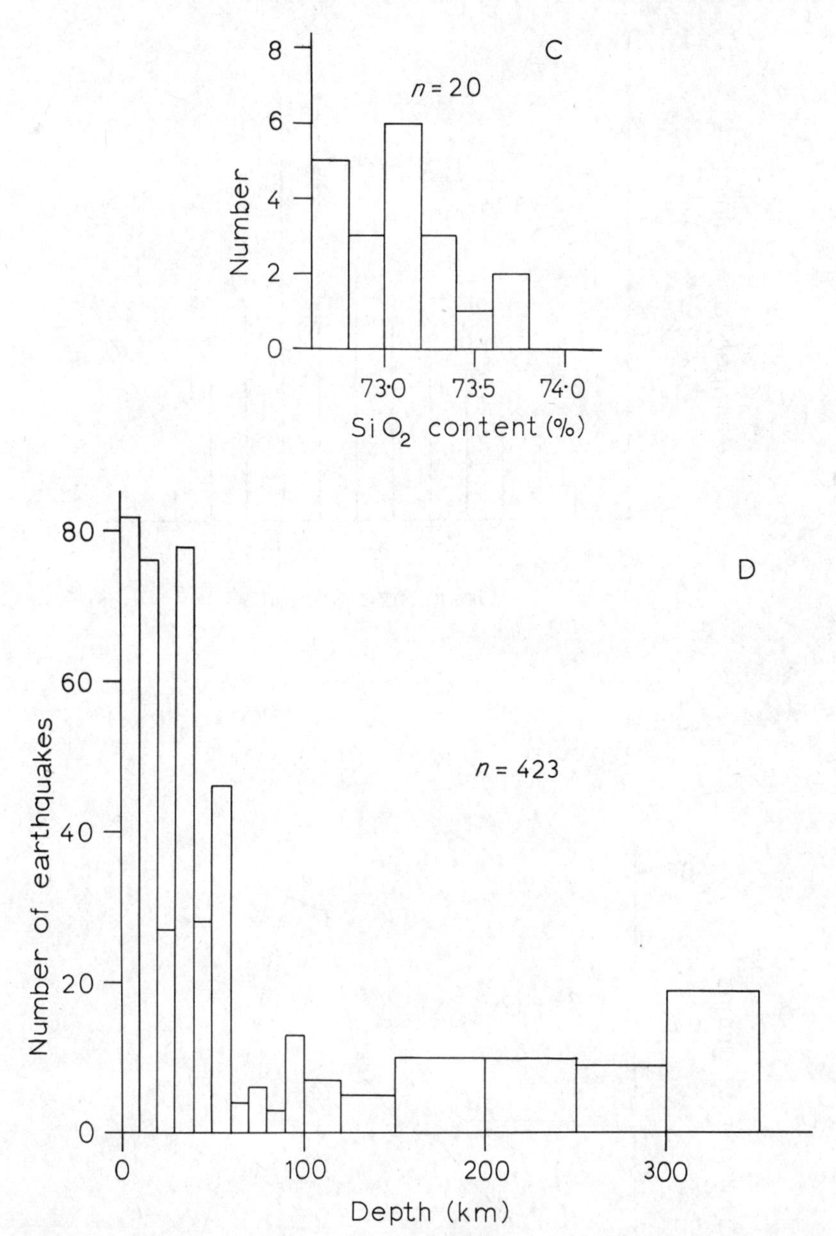

Figure 3.1 Histograms of various types of data: A—grain size analysis of a sediment (data from table 3.1); B—flow directions in pre-Cambrian rhyolites from the Wrekin, Shropshire (data provided by T. Prudence); C—replicate analyses of a sample of G − 1 for silica content (data provided by J. E. Thomas); D—number of earthquakes of magnitude greater than 6.0 at various depths beneath Japan during 1926–56 (data from Wadati, 1967).

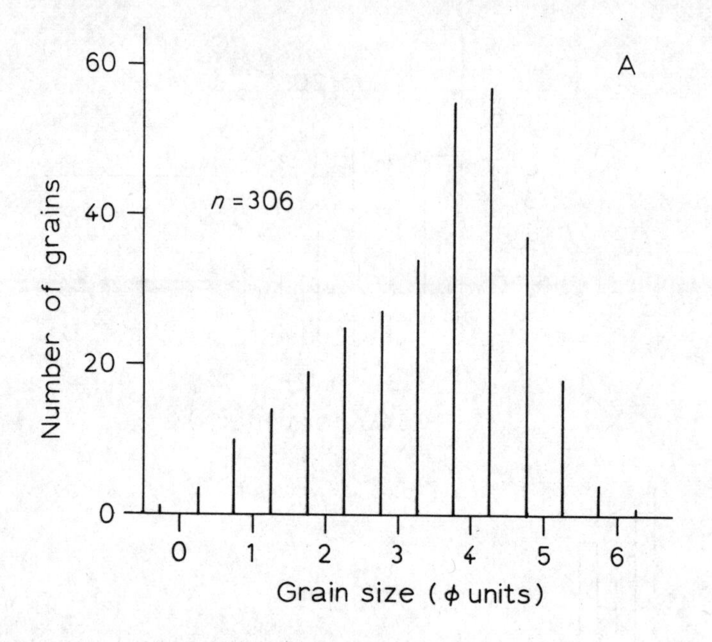
A
n = 306
Number of grains
0
20
40
60
0
1
2
3
4
5
6
Grain size (ϕ units)

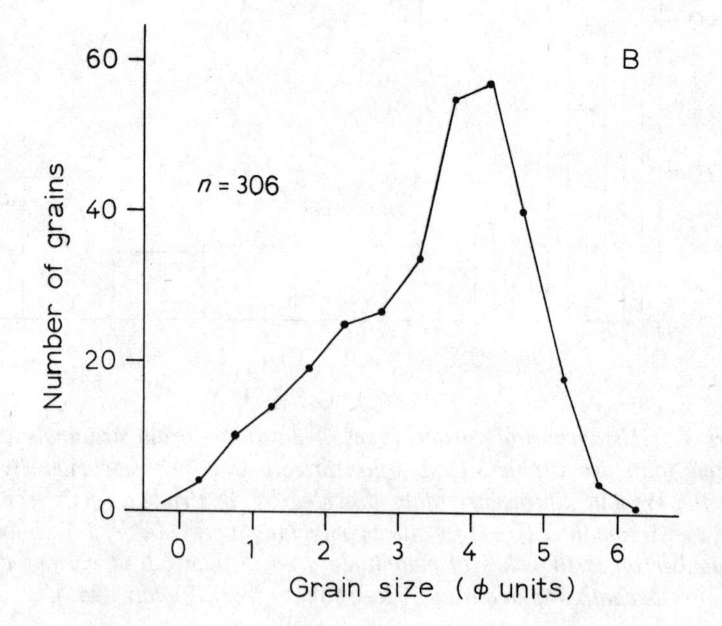
B
n = 306
Number of grains
0
20
40
60
0
1
2
3
4
5
6
Grain size (ϕ units)

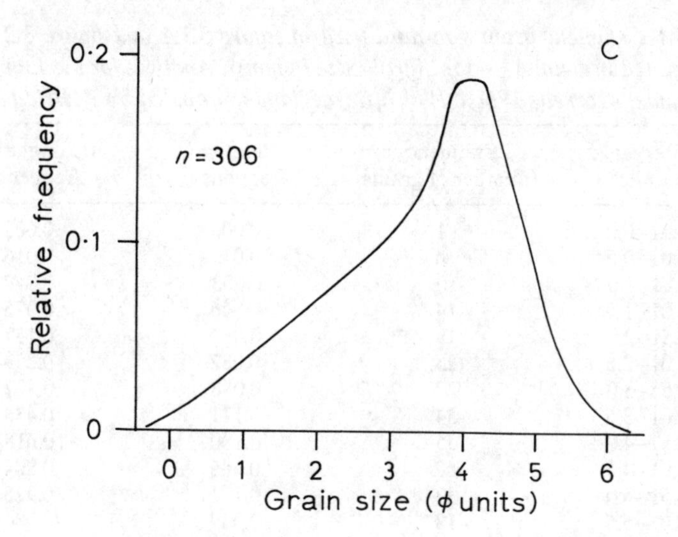

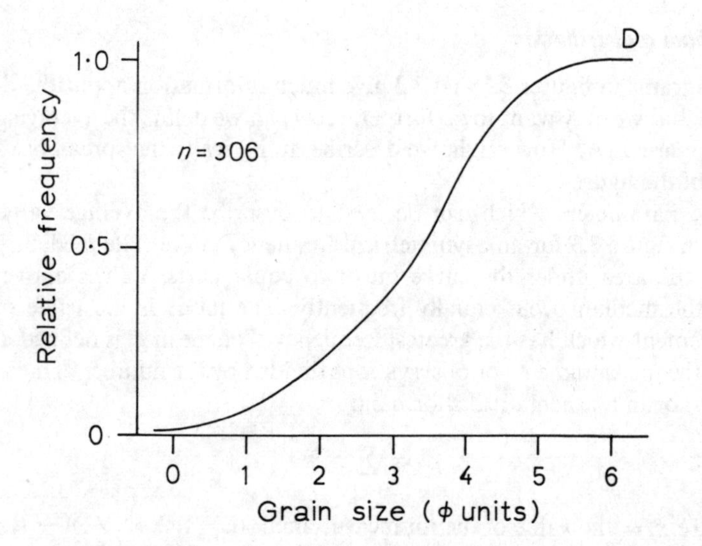

Figure 3.2 Some different ways of representing a frequency distribution, using the data from table 3.1: A—frequency chart; B—frequency polygon; C—relative frequency curve; D—cumulative frequency curve.

Table 3.1 Sediment grain size data used in figure 3.1A and figure 3.2. Grain size is quoted in ϕ units $\{-\log_2$ (grain size in mm)$\}$. Analysis for the Llandovery Sands, Morrells Wood, Welsh Borderlands, provided by P. Bridges.

Interval (ϕ units)	Frequency (number of grains)	Relative frequency	Cumulative frequency
−0.51–0.0	1	0.003	0.003
0.01–0.5	4	0.013	0.016
0.51–1.0	10	0.033	0.049
1.01–1.5	14	0.046	0.095
1.51–2.0	19	0.062	0.157
2.01–2.5	25	0.082	0.239
2.51–3.0	27	0.088	0.327
3.01–3.5	34	0.111	0.438
3.51–4.0	55	0.180	0.618
4.01–4.5	57	0.186	0.804
4.51–5.0	37	0.121	0.925
5.01–5.5	18	0.059	0.984
5.51–6.0	4	0.013	0.997
6.01–6.5	1	0.003	1.000

class mid-points. The cumulative frequency curve is simply produced in a similar fashion (figure 3.2D). (In the case of a sediment grain size curve it is customary to use normal probability paper, which is constructed from the normal distribution, as defined below).

Describing a distribution

The diagrams in figures 3.1 and 3.2 give much information about the distributions, but we may want to go further. How can we define the average grain size in figure 3.1A? How might we describe numerically the spread of values in any of the figures?

Three parameters which can be used to describe the average value are shown in figure 3.3 for an asymmetrical frequency curve. The median value divides the area under the curve into two equal parts. Values above and below the median occur equally frequently. The mode is the value of the measurement which has the greatest frequency. The mean μ, is defined as the sum of the measurements or observations divided by the number of measurements n; or in mathematical shorthand

$$\mu = \sum_{i=1}^{n} x_i/n \qquad (3.1)$$

where x_i = the value of the ith measurement ($i = 1, 2, 3, \ldots, n-1, n$)

We shall use $\sum x$ repeatedly. All it says is 'add together all the n values of x'

$$\sum_{i=1}^{n} x_i = x_1 + x_2 + x_3 + \ldots + x_n$$

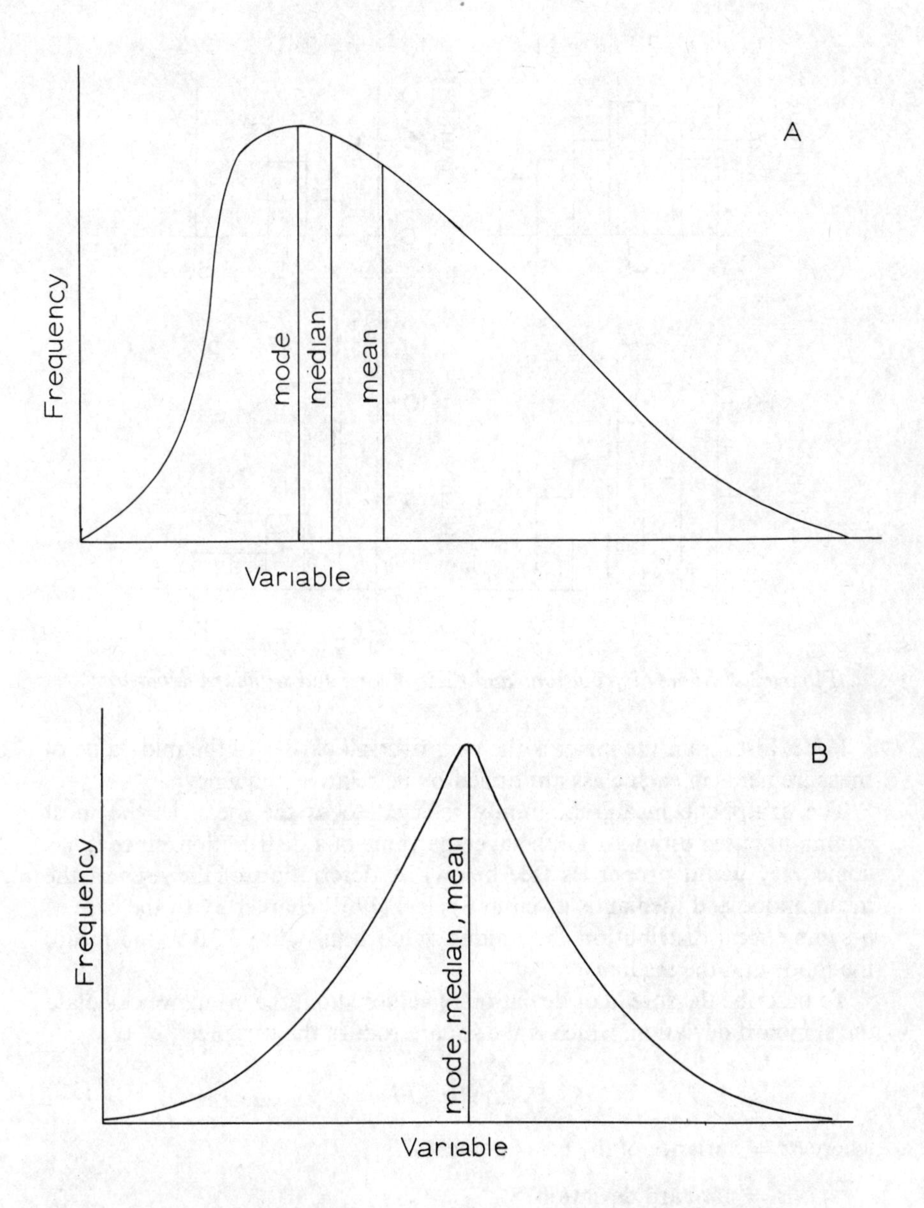

Figure 3.3 A—The median, mode and mean values of an asymmetrical frequency curve; B—The unique average value (which is median, mode and mean for a symmetrical frequency curve).

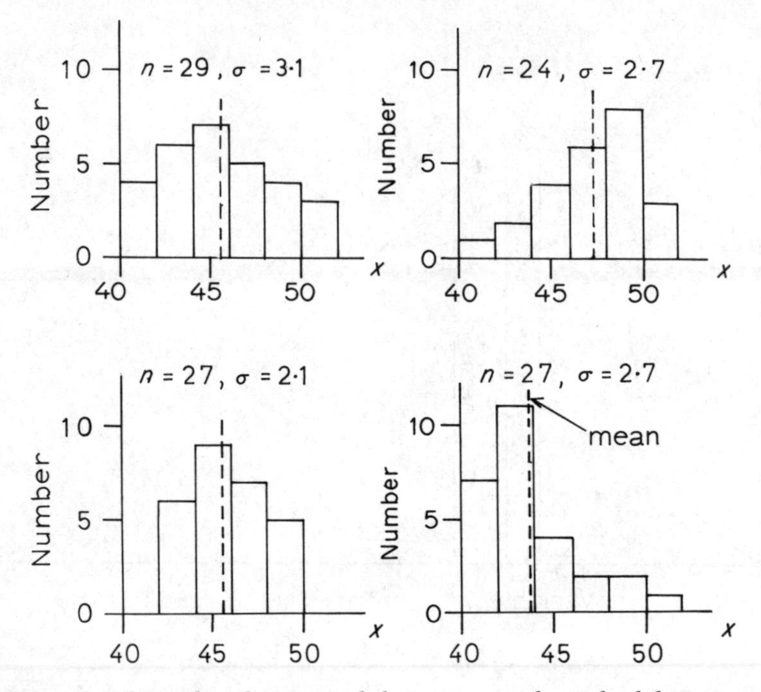

Figure 3.4 Some distributions and their means and standard deviations.

For a histogram the mean is the sum over all classes of the mid-value of measurements in each class multiplied by its relative frequency.

The arithmetic mean, commonly referred to as the mean, is the most commonly used estimator of the average value of a distribution, since it has some very useful properties (see below). A description of the uses of the mean, mode and median is given in Speigel (1961, chapter 3). In the case of a symmetrical distribution the unique value μ in figure 3.3B is the mean, the mode and the median.

To describe the spread or deviation of values about the mean, we calculate the standard deviation, which is the square root of the variance

$$\sigma^2 = \sum_{i=1}^{n} (x_i - \mu)^2/n \tag{3.2}$$

where σ^2 = variance of the measurements

σ = standard deviation

and the other symbols represent the same as they did in equation 3.1.

Empirically, σ seems a good measure, because it defines the sum of all the deviations of the measurements from the mean. Figure 3.4 shows that the larger the standard deviation the larger is the spread of values. The frequency curve is then wider and less clustered about the mean. The term $(x_i - \mu)^2$

is used for two reasons. Firstly it sums the total variation about the mean. In a symmetrical distribution the term $(x_i - \mu)$ would sum to zero since positive and negative deviations from the mean would cancel out. Secondly, subsequent statistical usage shows $(x_i - \mu)^2$ is more useful than the possible $(x_i - \mu)$ or modulus of $(x_i - \mu)$, which is the unsigned deviation from the mean.

These two properties of a curve, the mean and standard deviation, sometimes referred to as the first two moments, will be repeatedly used and manipulated in our statistical tests. The standard deviation is most useful for comparative purposes when derived from a symmetrical curve. Two further moments are sometimes used which reflect the skewness and the kurtosis or peakedness of a curve (see Miller and Kahn, 1962, or Krumbein and Pettijohn, 1938).

The shape of the distribution, or frequency density function, to give its full title, obtained from any set of measurements, usually approximates to one of a few defined forms, which are called probability density functions (p.d.f.).

This is very fortunate, since it allows us to quote a few well-known parameters (known as descriptive statistics) to define a distribution. These p.d.f. are related to each other and form the basis of our statistical tests. The remainder of this chapter, and the following chapter, deal with the derivation and properties of these distributions which will be used repeatedly in subsequent chapters, as the basis of many statistical tests.

3.2 The binomial distribution

We can derive the form of the binomial distribution by considering a simple geological example and using the probability formulae we developed in chapter 2.

Imagine we have the contents of a very large bag of sand spread evenly on a tray. In this sand a proportion P are grains of apatite distributed evenly throughout the population. This leaves a proportion Q (equal to $[1 - P]$) that are non-apatite grains (they would probably be quartz, zircon, tourmaline, etc.). We will try to estimate P by taking a few grains from the tray and identifying them. If we choose these few grains (N in number) so that every grain in the tray has an equal chance of being taken, we say that we are estimating P by taking a random sample (see chapter 4). We then repeat this sampling procedure T times and try the whole operation for different sizes of samples (that is, different numbers of grains).

1. One-grain samples If we select one sand grain randomly from the tray, the probability of it being a grain of apatite is P, and of it being a non-apatite is Q (equal to $[1 - P]$). Obviously if we take only one grain then the sample is either apatite or non-apatite! If we do this T times we would expect to get

Table 3.2 One-grain samples ($N = 1$).

Expected number of samples in which	
Grain is a non-apatite (0% apatite)	Grain is an apatite (100% apatite)
$TQ(=T\{1 - P\})$	TP

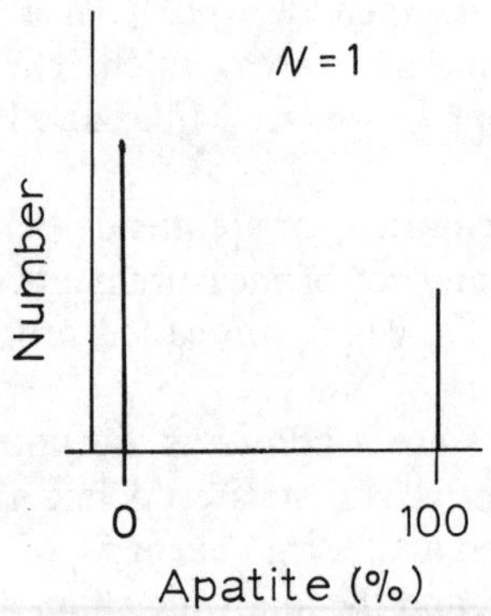

Figure 3.5 Frequency distribution for one-grain samples ($N = 1$).

an apatite grain TP times and a non-apatite grain TQ times (equal to $T[1 - P]$ times)

Figure 3.5 shows the frequency distribution, which is not terribly inspiring, giving us two possible estimates of P.

2. Two-grain samples Again if we select our two grains randomly we can build on the one-grain case. The first of the two grains will be non-apatite in TQ cases. The second grain will be a non-apatite in Q of these. So $TQ \times Q$ of the two-grain samples will be both non-apatites. Similarly, we can consider the other cases (using the probability rules of chapter 2).

Table 3.3 T Samples ($N = 2$).

First grain is a non-apatite TQ of these		First grain is an apatite TP of these	
Second grain is a non-apatite	Second grain is an apatite	Second grain is a non-apatite	Second grain is an apatite
TQ^2	TQP	TPQ	TP^2

This gives us three estimates of P, from our two-grain samples which can be all non-apatite, half apatite or all-apatite grains

Table 3.4 Two-grain samples ($N = 2$).

Expected number of samples in which there are		
No apatites (0% apatite)	One apatite (50% apatite)	Both apatite (100% apatite)
TQ^2	$2TQP$	TP^2

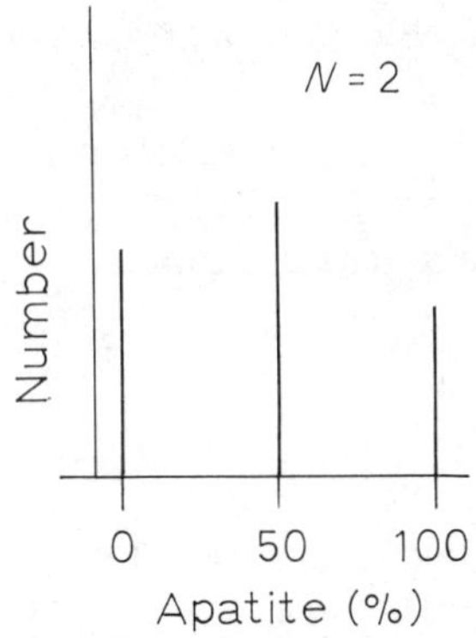

Figure 3.6 Frequency distribution for two-grain samples ($N = 2$).

and the frequency distribution of figure 3.6 looks better, giving us a closer estimate of p.

3. Three-grain samples Here we have four possible outcomes for a random sample and our table can be built up just as before

Table 3.5 Three grain samples ($N = 3$).

Expected number of samples in which there are			
No apatites (0% apatite)	One apatite (33.3%)	Two apatites (66.7%)	All apatites (100% apatite)
TQ^3	$3TQ^2P$	$3TQP^2$	TP^3

which produces the frequency distribution in figure 3.7.

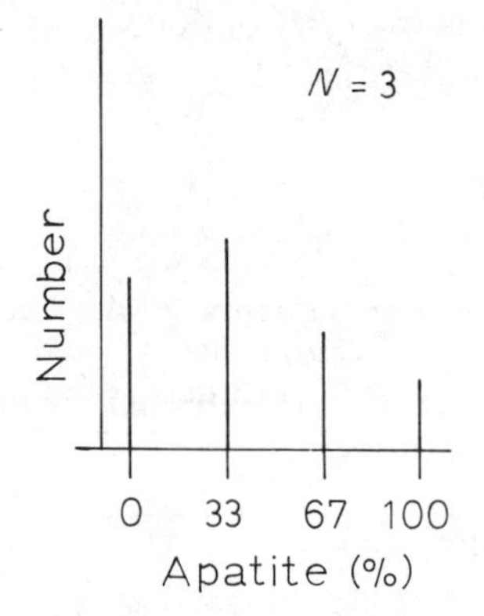

Figure 3.7 Frequency distribution for three-grain samples ($N = 3$).

We could extend this treatment to successively larger samples (figure 3.8 shows the distribution for ten-grain samples), but we have sufficient information already to deduce a general formula. Summarising the frequencies we obtained

When $N = 1$: $TQ; TP$

$N = 2$: $TQ^2; 2TQP; TP^2$

$N = 3$: $TQ^3; 3TQ^2P; 3TQP^2; TP^3$

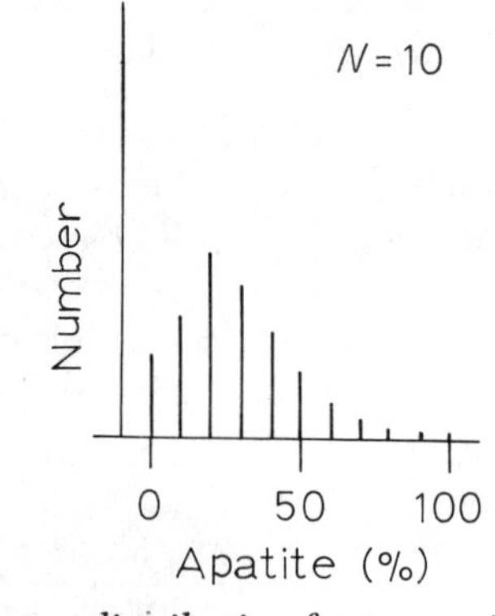

Figure 3.8 Frequency distribution for ten-grain samples ($N = 10$).

These are the terms of the expansion of

$N = 1$: $T(Q + P)^1$

$N = 2$: $T(Q + P)^2$

$N = 3$: $T(Q + P)^3$

which in the general case is the binomial theorem $T(Q + P)^N$ with terms

$$TQ^N; TNQ^{N-1}P^1; \frac{TN(N-1)Q^{N-2}P^2}{1 \times 2}; \frac{TN(N-1)(N-2)Q^{N-3}P^3}{1 \times 2 \times 3}; \ldots$$

$$\ldots TN\frac{(N-1)(N-2)\ldots 2Q^1P^{N-1}}{1 \times 2 \times 3 \ldots \times (N-1)}; TP^N$$

if we use factorials ($F! = 1 \times 2 \times 3 \ldots \times F$) the Rth term of this expansion can be written

$$\frac{TN!}{R!(N-R)!}Q^{N-R}P^R$$

which leads to a general statement about the form of the binomial distribution. This discrete (that is, discontinuous) probability density function can be used when a given event, or trial, has two possible outcomes. These outcomes may, for example, be apatite or non-apatite in a sand, or presence or absence of a fossil.

If p is the probability that an event will occur in any given trial (and $q = 1 - p$ is the probability of that same event not occurring), then the probability $p(x)$ that the event will happen x times in a trial where n items are picked is

$$p(x) = \binom{n}{x} p^x q^{n-x} = \frac{n!}{x!(n-x)!} p^x q^{n-x} \tag{3.3}$$

If we repeat our trial m times then the event should occur $mp(x)$ times. In our apatite example, if we take a single three-grain sample ($n = 3$) from a sediment which is 5 per cent apatite ($p = 0.05$), the probability of having two apatite grains ($x = 2$) in the sample is

$$p(x) = \frac{3!}{2!1!} p^2 q^1 = 3p^2 q = 3 \times (0.05)^2 \times 0.95$$

$$= 0.0071$$

From the general formula we can produce the full discrete binomial distribution by substituting values of x from zero to n. This is the distribution of the sample sum. Figure 3.9 shows the form of this distribution for a number of cases. It has the properties

mean value $= np$

variance $= npq$

standard deviation $= \sqrt{(npq)}$

The binomial distribution is not greatly used in geology, but it is an important base from which to derive other distributions. Kahn (1956) gives

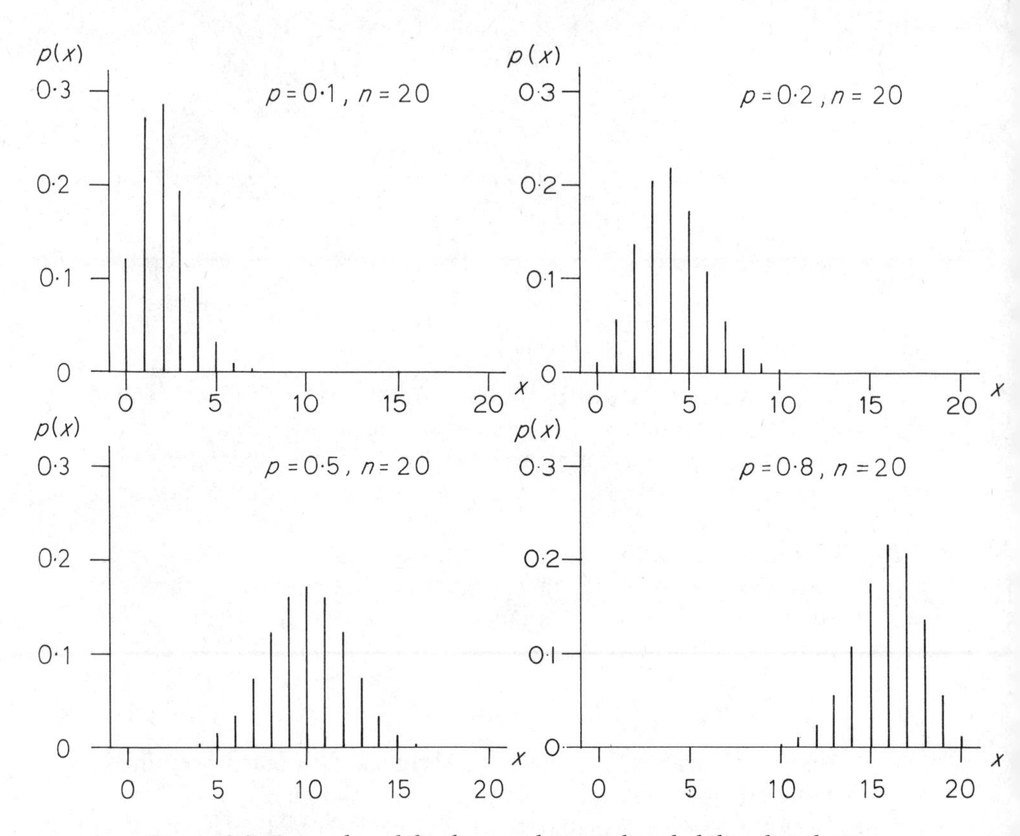

Figure 3.9 Examples of the discrete binomial probability distribution.

an example of its application to the probability of grain-to-grain contacts in thin-section work. Chayes (1956) applies it to petrographic modal analysis and Yevjevich (1971) gives examples of its use in hydrology.

3.3 The normal distribution and its standardised form

Figure 3.10A shows a discrete binomial distribution for a large value of n. We can imagine that as the size of our sample becomes infinite this frequency chart gradually becomes a complete area bounded by a continuous curve stretching infinitely in either direction (from $-\infty$ to $+\infty$). Intuitively this is easy to visualise, and in a simple sentence we have produced the extremely important normal distribution shown in figure 3.10B.

We are able to prove this mathematically by considering equation 3.3 for very large values of n and using an approximation for calculating very large factorials (Stirling's approximation). It involves a page of algebra

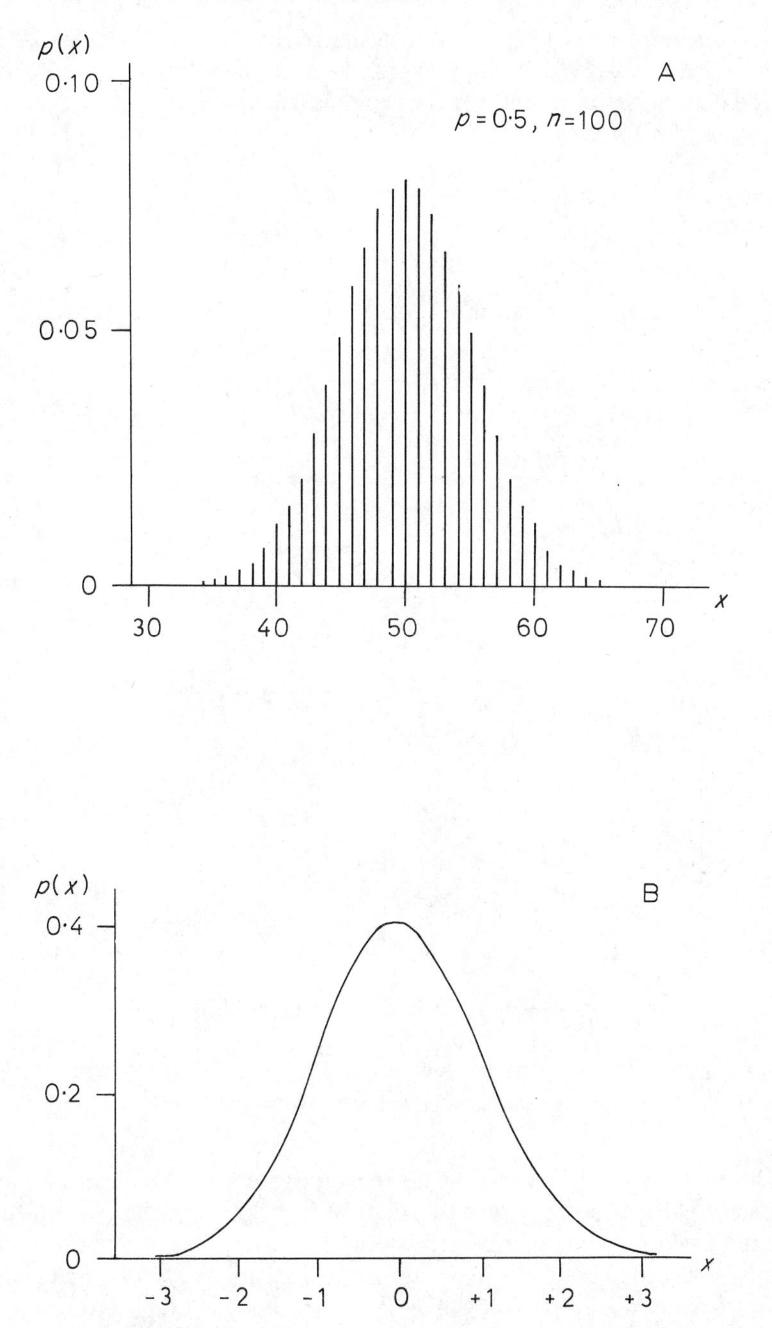

Figure 3.10 A—Frequency polygon for the discrete binomial distribution for a large value of n; B—The continuous normal or gaussian probability distribution.

which does not need to be reproduced here. Barford (1967, p. 86–7) and Yule and Kendall (1953, p. 177–81) give the proof in full.

The continuous probability distribution is called the normal or Gaussian distribution defined by

$$y = \frac{1}{\sigma\sqrt{(2\pi)}} \exp\left[-\tfrac{1}{2}(x - \mu)^2/\sigma^2\right] \qquad (3.4)$$

where μ = mean value

σ = standard deviation

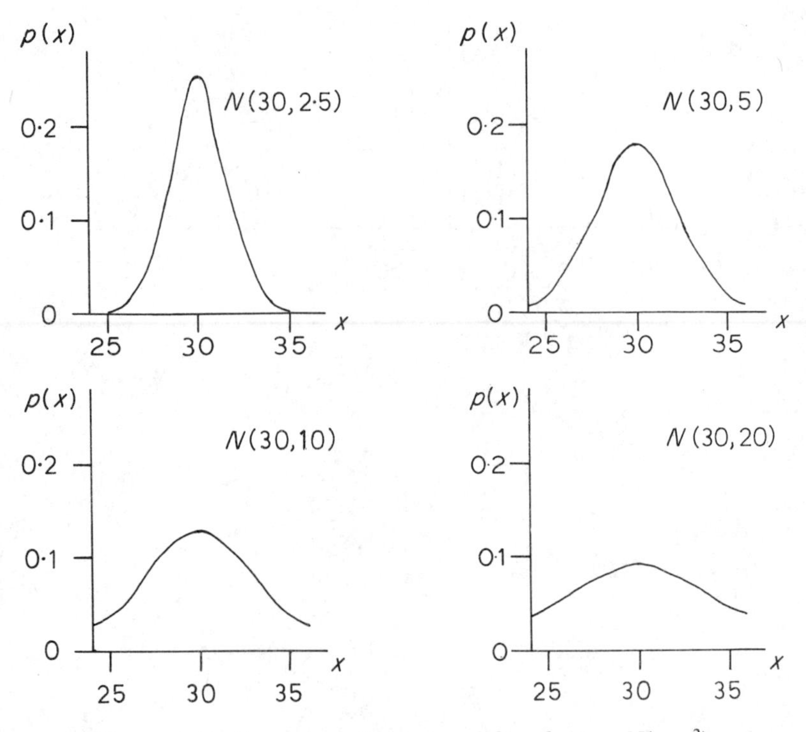

Figure 3.11 Examples of the normal distribution, $N(\mu, \sigma^2)$.

Its very characteristic bell shape is seen in figure 3.11. We often refer to a given normal distribution by the shorthand $N(\mu, \sigma^2)$. Thus we have completely defined its form. This distribution is very important because many populations are distributed in this way, be it measurements of intelligence, measurements of the porosity of a sandstone body or replicate chemical analyses.

The total area under this curve is one square unit. Hence the area under the curve between two x values a and b gives the proportion of x values that lie between a and b. This proportion tells us how likely we are to get an x value that will lie between a and b—that is the probability of an x value between a and b occurring. So we can use the area under our normal curve directly to tell us the probability of an x value occurring in the specified range, a to b. This is important because once we know a given measurement is normally distributed, with a given mean and standard deviation, we can precisely define the probability (or likelihood) of occurrence of a given value or range of values. We calculate the area

$$p[a \leqslant x \leqslant b] = \text{area under curve} = \int_a^b \frac{1}{\sigma\sqrt{(2\pi)}} \exp\left[-\tfrac{1}{2}(x-\mu)^2/\sigma^2\right] \mathrm{d}x$$

In practice we do not try and solve this each time, because areas under the normal curve are tabulated in terms of the number of standard deviations from the mean, as shown in table 3.6. The variable z is defined as $(x - \mu)/\sigma$, where x is the value being considered. These tables only quote positive deviations from the mean, since the whole curve is symmetrical about that point.

As an example, suppose a sandstone body has an average porosity of 20 per cent void space, distributed normally, with a standard deviation of 2 per cent, as shown in figure 3.12. What is the probability of a given specimen having a porosity between 17.5 and 23 per cent void space? To answer this we must find the area under the curve between these values

$$x = 17.5, \text{ gives } (x - \mu)/\sigma = (17.5 - 20)/2 = -1.25$$

$$x = 23, \text{ gives } (x - \mu)/\sigma = (23 - 20)/2 = +1.5$$

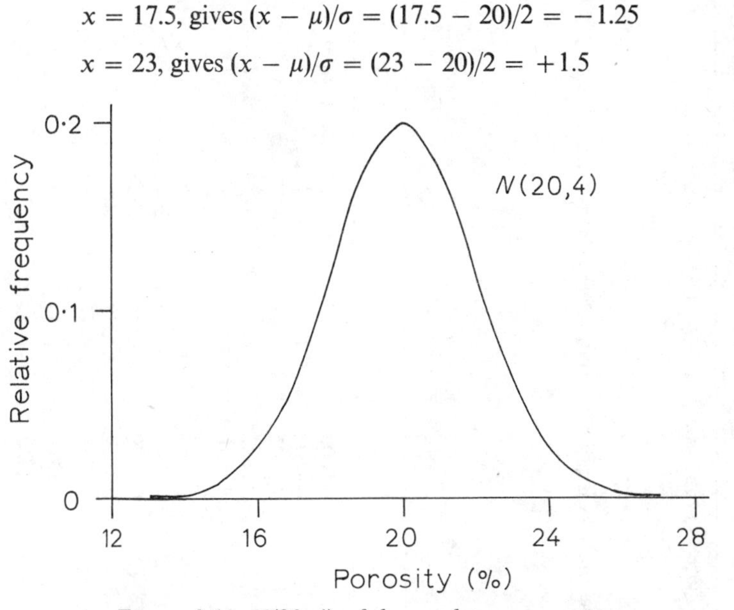

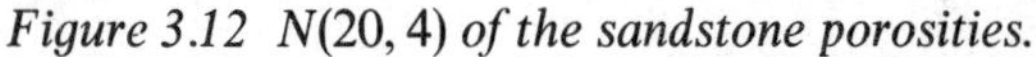

Figure 3.12 N(20, 4) *of the sandstone porosities.*

Table 3.6 Areas under the normal curve, tabulated by z values. Table taken from Murdoch and Barnes (1974, table 3), with permission.

The function tabulated is $1 - \phi(z)$ where $\phi(z)$ is the cumulative distribution function of a standardised normal variable z. Thus

$$1 - \phi(z) = \frac{1}{\sqrt{(2\pi)}} \int_z^\infty \exp(-\tfrac{1}{2}x^2)\,dx$$

is the probability that a standardised normal variable selected at random will be greater than a value of $z = (x - \mu)/\sigma$.

$(x - \mu)/\sigma$	0.00	0.01	0.02	0.03	0.04	0.05	0.06	0.07	0.08	0.09
0.0	0.5000	0.4960	0.4920	0.4880	0.4840	0.4801	0.4761	0.4721	0.4681	0.4641
0.1	0.4602	0.4562	0.4522	0.4483	0.4443	0.4404	0.4364	0.4325	0.4286	0.4247
0.2	0.4207	0.4168	0.4129	0.4090	0.4052	0.4013	0.3974	0.3936	0.3897	0.3859
0.3	0.3821	0.3783	0.3745	0.3707	0.3669	0.3632	0.3594	0.3557	0.3520	0.3483
0.4	0.3446	0.3409	0.3372	0.3336	0.3300	0.3264	0.3228	0.3192	0.3156	0.3121
0.5	0.3085	0.3050	0.3015	0.2981	0.2946	0.2912	0.2877	0.2843	0.2810	0.2776
0.6	0.2743	0.2709	0.2676	0.2643	0.2611	0.2578	0.2546	0.2514	0.2483	0.2451
0.7	0.2420	0.2389	0.2358	0.2327	0.2296	0.2266	0.2236	0.2206	0.2177	0.2148
0.8	0.2119	0.2090	0.2061	0.2033	0.2005	0.1977	0.1949	0.1922	0.1894	0.1867
0.9	0.1841	0.1814	0.1788	0.1762	0.1736	0.1711	0.1685	0.1660	0.1635	0.1611
1.0	0.1587	0.1562	0.1539	0.1515	0.1492	0.1469	0.1446	0.1423	0.1401	0.1379
1.1	0.1357	0.1335	0.1314	0.1292	0.1271	0.1251	0.1230	0.1210	0.1190	0.1170
1.2	0.1151	0.1131	0.1112	0.1093	0.1075	0.1056	0.1038	0.1020	0.1003	0·0985

1.3	0·0968	0.0951	0.0934	0.0918	0.0901	0.0885	0.0869	0.0853	0·0838	0·0823
1.4	0.0808	0.0793	0.0778	0.0764	0.0749	0.0735	0.0721	0.0708	0.0694	0.0681
1.5	0.0668	0.0655	0.0643	0.0630	0.0618	0.0606	0.0594	0.0582	0.0571	0.0559
1.6	0.0548	0.0537	0.0526	0.0516	0.0505	0.0495	0.0485	0.0475	0.0465	0.0455
1.7	0.0446	0.0436	0.0427	0.0418	0.0409	0.0401	0.0392	0.0384	0.0375	0.0367
1.8	0.0359	0.0351	0.0344	0.0336	0.0329	0.0322	0.0314	0.0307	0.0301	0.0294
1.9	0.0287	0.0281	0.0274	0.0268	0.0262	0.0256	0.0250	0.0244	0.0239	0.0233
2.0	0.02275	0.02222	0.02169	0.02118	0.02068	0.02018	0.01970	0.01923	0.01876	0.01831
2.1	0.01786	0.01743	0.01700	0.01659	0.01618	0.01578	0.01539	0.01500	0.01463	0.01426
2.2	0.01390	0.01355	0.01321	0.01287	0.01255	0.01222	0.01191	0.01160	0.01130	0.01101
2.3	0.01072	0.01044	0.01017	0.00990	0.00964	0.00939	0.00914	0.00889	0.00866	0.00842
2.4	0.00820	0.00798	0.00776	0.00755	0.00734	0.00714	0.00695	0.00676	0.00657	0.00639
2.5	0.00621	0.00604	0.00587	0.00570	0.00554	0.00539	0.00523	0.00508	0.00494	0.00480
2.6	0.00466	0.00453	0.00440	0.00427	0.00415	0.00402	0.00391	0.00379	0.00368	0.00357
2.7	0.00347	0.00336	0.00326	0.00317	0.00307	0.00298	0.00289	0.00280	0.00272	0.00264
2.8	0.00256	0.00248	0.00240	0.00233	0.00226	0.00219	0.00212	0.00205	0.00199	0.00193
2.9	0.00187	0.00181	0.00175	0.00169	0.00164	0.00159	0.00154	0.00149	0.00144	0.00139
3.0	0.00135									

The area under the normal curve between

$$\text{mean } (z = 0) \text{ and } +1.5 \quad = 0.4332$$

$$\text{mean } (z = 0) \text{ and } -1.25 = 0.3944$$

Thus the total area between $z = +1.5$ and $-1.25 = 0.8276$.

Therefore the probability of a porosity value between 17.5 and 20 per cent for our sandstone $N(20, 4)$ is 0.8276. So 82.76 per cent of possible determined values would be between these two measurements. (We do make a slight approximation, since the normal distribution really continues from $-\infty$ to $+\infty$, which the porosity values cannot do).

Table 3.6 shows that 2.5 per cent of the area of our curve lies beyond $z = +1.96$, and similarly 2.5 per cent of the curve lies beyond $z = -1.96$. These small areas in each tail of the distribution are referred to as the critical regions (see next chapter).

If

$$z + 1.96 = (x - \mu)/\sigma = (x - 20)/2 \text{ then } x = 23.92$$

If

$$z - 1.96 = (x - \mu)/\sigma = (x - 20)/2 \text{ then } x = 16.08$$

For the porosity values, 95 per cent of the curve, placed symmetrically about the mean, lies between 23.92 and 16.08 per cent void space. Directly from table 3.6, $\mu \pm 1.96\sigma$ is 20 ± 3.92 per cent. Likewise 99 per cent of the area lies between $\mu \pm 2.576\sigma$. These two area percentages, or probabilities will be very important in later chapters. Written in shorthand

$$p_{0.95} = \mu \pm 1.96\sigma$$

$$p_{0.99} = \mu \pm 2.576\sigma$$

for $N(\mu, \sigma^2)$

In addition we should note that $\mu \pm \sigma$ encloses 68.27 per cent of the area under the curve, $\mu \pm 2\sigma$ encloses 95.45 per cent and $\mu \pm 3\sigma$ encloses 99.73 per cent.

The standardised normal distribution

We have used z to express deviations from the mean of the normal distribution in terms of standard deviations. If we transform any given normal distribution by re-expressing the formula for $N(\mu, \sigma^2)$ in terms of z, we obtain a new normal distribution

$$y = \frac{1}{\sqrt{(2\pi)}} \exp\left[-\tfrac{1}{2}z^2\right] \qquad (3.5)$$

which has a mean of zero and a standard deviation of unity, and of course an area of one square unit. This is $N(0, 1)$ the standardised normal distribution (figure 3.13). It is very useful to use. For example, between $z = \pm 1$, we find an area of 0.6827, and 95 per cent of the curve lies between $z = \pm 1.96$. Our table 3.6 is now a direct listing of areas under the standardised normal distribution between zero (the mean) and a given ordinate.

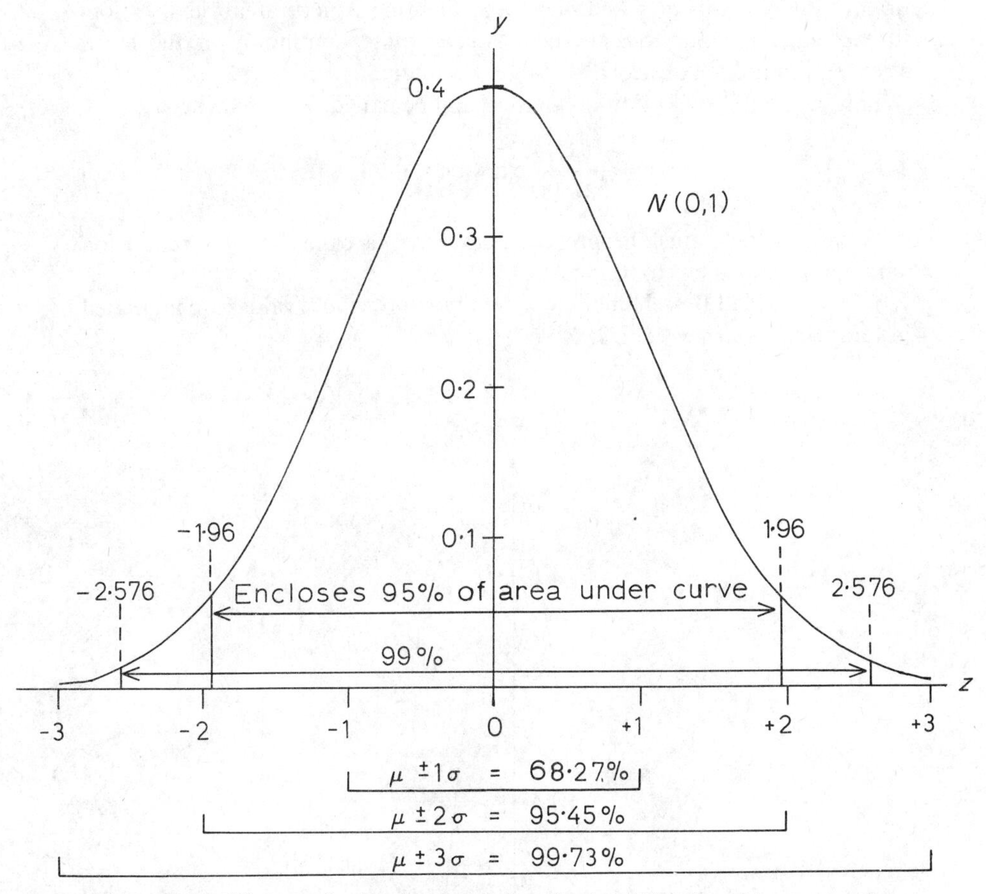

Figure 3.13 The standardised normal distribution N(0, 1) and its properties.

The ideas developed in these last two sections will be employed repeatedly. The use of areas under the distribution of a variable, to define the probability of a given value of that variable occurring, is basic to all statistical testing. To really understand this it is well worth generating distributions from random numbers on the computer, as described in Till *et al.* (1972). In addition, all students should have their own sets of data, collected in the laboratory or in the field, which can be graphed to study the normal distribution. Krumbein and Graybill (1965) tabulate a number of types of data which would be suitable.

3.4 Some other distributions

The circular normal or von Mises distribution

This is the equivalent of the normal distribution for a population of directional measurements. These could be slopes, cross-bedding directions, orientation of fossils on a bedding plane, or some structural fabric direction. In fact such angular measurements occur quite commonly in the earth sciences, but are not always distributed normally.

This 'normal distribution on a circle' can be called $M(\mu_0, \kappa)$ where

$$y = \frac{1}{2\pi I_0(\kappa)} \exp\{\kappa \cos(\theta - \mu_0)\} \tag{3.6}$$

κ (kappa), which must be greater than zero, is called the concentration parameter (equivalent to σ).

$I_0(\kappa)$ is a modified Bessel function of the first kind, whose values are tabulated in Gray and Matthews (1922, table VII).

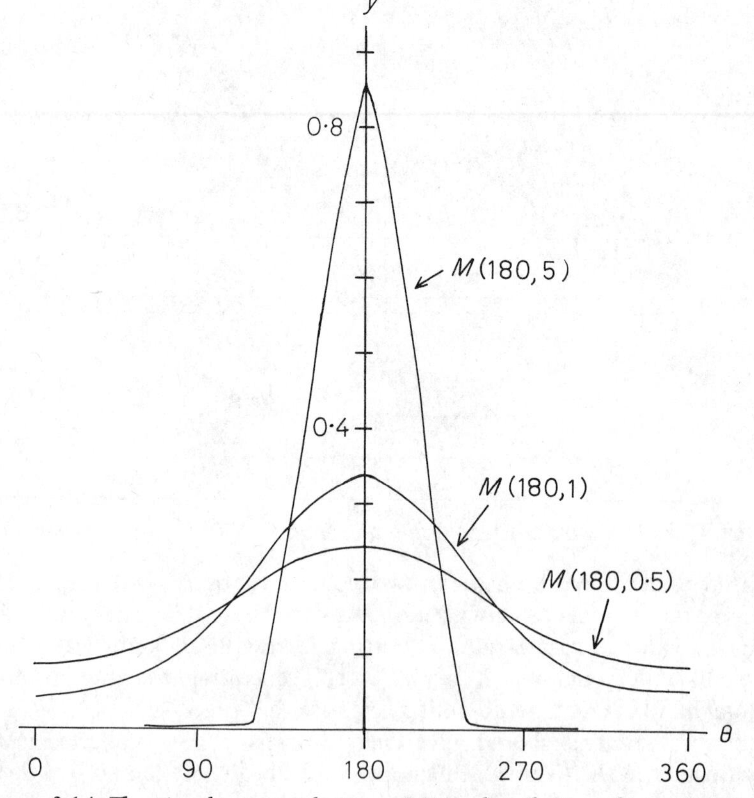

Figure 3.14 The circular normal or von Mises distribution, for various values of κ. θ is in degrees.

θ and μ_0 lie between 0 and 2π radians (that is 0° and 360°) and are the measurement and the mean of the measurements respectively.

The distribution is plotted in figure 3.14, and looks like a normal distribution, but the ordinate is in angular measures.

Table 3.7 gives a set of angular measurements, which are plotted in figure 3.15. If we have n measurements of angles θ_i ($i = 1, 2, 3, \ldots n$), then their co-ordinates on a unit circle are

$$x_i = \cos\theta_i \quad \text{and} \quad y_i = \sin\theta_i$$

and their means are

$$\bar{x} = \sum_{i=1}^{n} \cos\theta_i/n$$

$$\bar{y} = \sum_{i=1}^{n} \sin\theta_i/n$$

If we represent this mean point in polar co-ordinates $(r, \bar{\theta})$ as in figure 3.16

$$r = \sqrt{(\bar{x}^2 + \bar{y}^2)}$$

$$\cos\bar{\theta} = \bar{x}/r, \quad \sin\bar{\theta} = \bar{y}/r$$

Table 3.7 Directional data plotted in figures 3.15 and 3.16. Measurements of palaeocurrent azimuths from the Jura Quartzite, Islay. Data provided by R. Anderton.

Direction of azimuth (°E of N)	Sine of angle (y_i)	Cosine of angle (x_i)
12	+0.2079	+0.9781
353	−0.1392	+0.9903
359	−0.0175	+0.9998
332	−0.4695	+0.8829
341	−0.3256	+0.9455
299	−0.8746	+0.4848
30	+0.5000	+0.8660
24	+0.4067	+0.9135
53	+0.7986	+0.6018
284	−0.9703	+0.2419
99	+0.9877	−0.1564
72	+0.9511	+0.3090
28	+0.4695	+0.8829
93	+0.9986	−0.0523
125	+0.8192	−0.5736
318	−0.6691	+0.7431
3	+0.0524	+0.9986
45	+0.7071	+0.7071
Sum	+3.4330	+10.7630
Mean	+0.1907	+0.5979

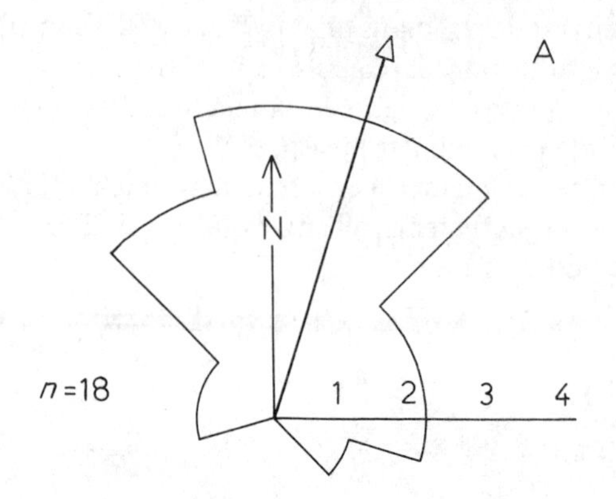

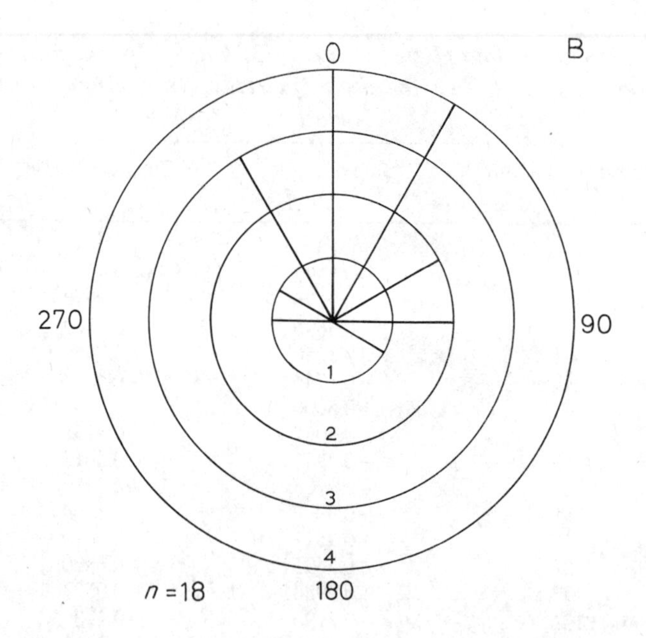

Figure 3.15 Two ways of plotting the directional data of palaeocurrent azimuths given in table 3.7: A—Rose diagram, using frequency of angles in 30° classes (345–015, 015–045, etc.); B—Circular frequency chart with the same classes as figure 3.15 A.

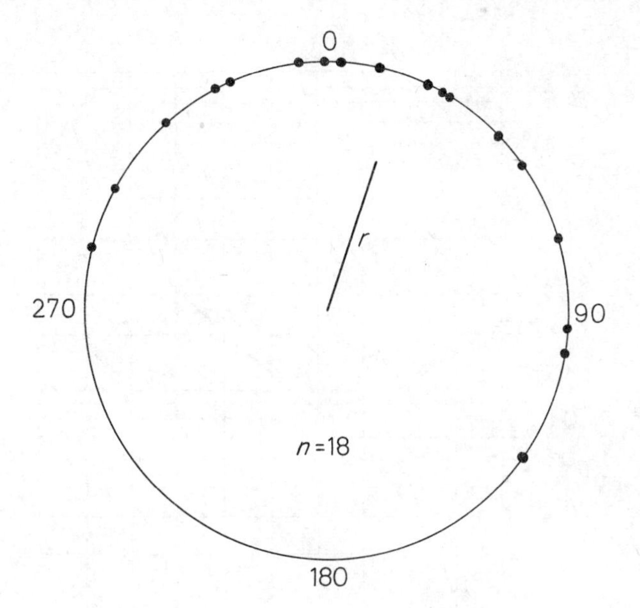

Figure 3.16 The directional data from table 3.7 and its mean vector plotted on the unit circle.

Now this $\bar{\theta}$ is the μ_0 of the von Mises distribution, but r is not its concentration parameter, κ. The value of r is an estimate of the spread of the angular values around a unit circle. The closer the points are clustered, the nearer is r to unity. The mean angular deviation s, which is like a standard deviation, can be calculated

$$s = \sqrt{[2(1 - r)]}$$

κ must be looked up in tables from the value of r (Batschelet, 1965; Mardia, 1972), or κ can be interpolated less accurately from figure 3.17.

For our example

$$\bar{x} = +\,0.5979$$

$$\bar{y} = +\,0.1907$$

$$r = [(0.5979)^2 + (0.1907)^2]^{\frac{1}{2}} = 0.6276$$

$$\cos\bar{\theta} = 0.9527$$

$$\sin\bar{\theta} = 0.3039$$

$$\mu_0 = \bar{\theta} = 18^\circ = 0.314 \text{ radians}$$

$$s = [2(1 - 0.6276)]^{\frac{1}{2}} = 0.863 \text{ radians} = 60^\circ$$

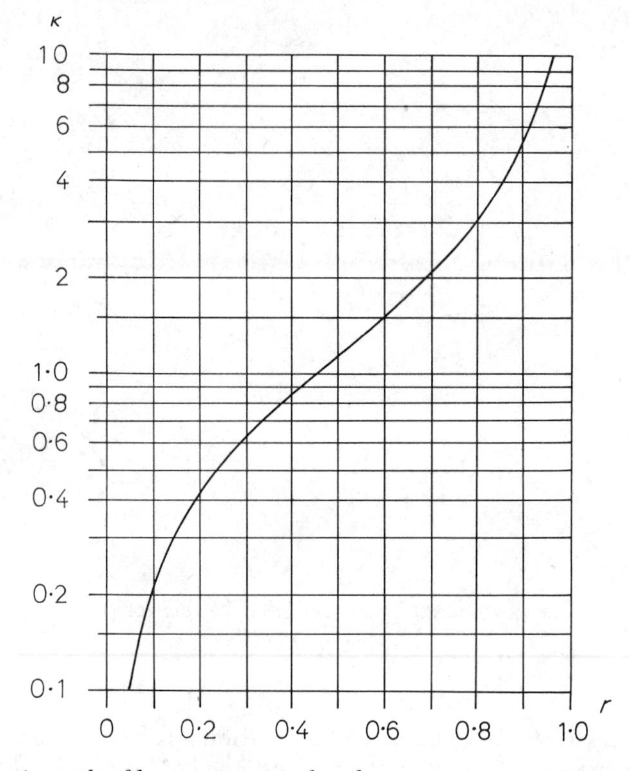

Figure 3.17 A graph of best estimates of κ, the concentration parameter, from r, the mean vector, for von Mises distributions. Values taken from Mardia (1972, Appendix 2.2).

and from figure 3.17

$$\kappa = 1.65$$

giving a distribution

$$y = \frac{1}{2\pi I_0(1.65)} \exp\left[1.65 \cos\left(\theta - 0.314\right)\right]$$

The mean vector has a direction of 18° and the value of s tells us that about two thirds of the angular measurements lie within an angle of $18° \pm 60°$.

Further details about the use of statistics for directional data can be found in Watson (1966) and Reyment (1971). If you can cope with a full mathematical treatment of all sorts of directional data, from cross bedding in sands, to the vanishing angle of British mallards, see Mardia (1972) who also deals fully with spherical distributions. A whole series of tests, equivalent to t-tests F-tests and χ^2 tests are available for directional data. These allow comparisons to be made between sets of directional data, and operate like those for the normal distribution described in the following chapters. Mardia (1972) describes and gives us worked examples for all of them.

The log-normal distribution

In the 1950s and 1960s a large amount of trace-element data was being produced for rocks using spectrographic methods. One of the proponents of the technique noticed that such data sets had a non-normal distribution (Ahrens, 1954). Figure 3.18 shows the type of distribution Ahrens would have obtained. This is called a log-normal distribution because if the logarithms of these trace element concentrations are plotted a normal distribution results.

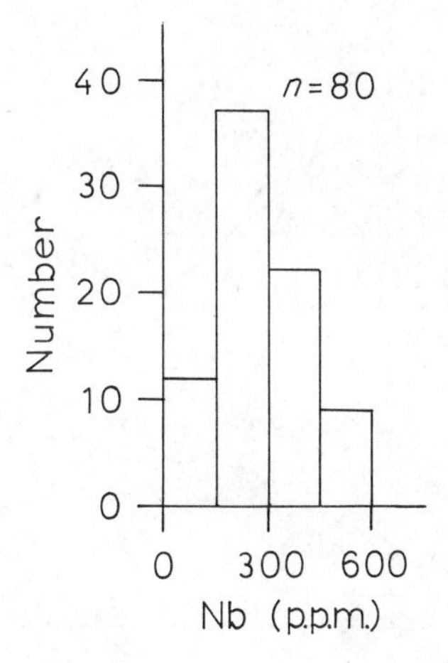

Figure 3.18 Niobium content of peralkaline oversaturated lavas. Data provided by D. K. Bailey.

The equation

$$y = \frac{1}{x\sigma_n\sqrt{(2\pi)}}\exp\left[-\frac{1}{2}\left(\frac{\ln x - \mu_n}{\sigma_n}\right)^2\right] \tag{3.7}$$

describes the log-normal distribution $L(\mu_n, \sigma_n)$ where

μ_n = mean value of ln x's (natural log of x's)

σ_n = standard deviation of ln x's

The mean μ, and standard deviation σ, of the actual x values can be calculated

$$\mu = \exp(\mu_n + \tfrac{1}{2}\sigma_n^2)$$

$$\sigma^2 = \mu^2[\exp(\sigma_n^2) - 1]$$

A few examples of the log-normal distribution are shown in figure 3.19. Looking in other texts we find some confusion in the symbols used for μ_n and μ, and σ_n and σ. For clarity, the form of Aitchison and Brown (1957), who wrote a textbook about the log-normal distribution has been followed.

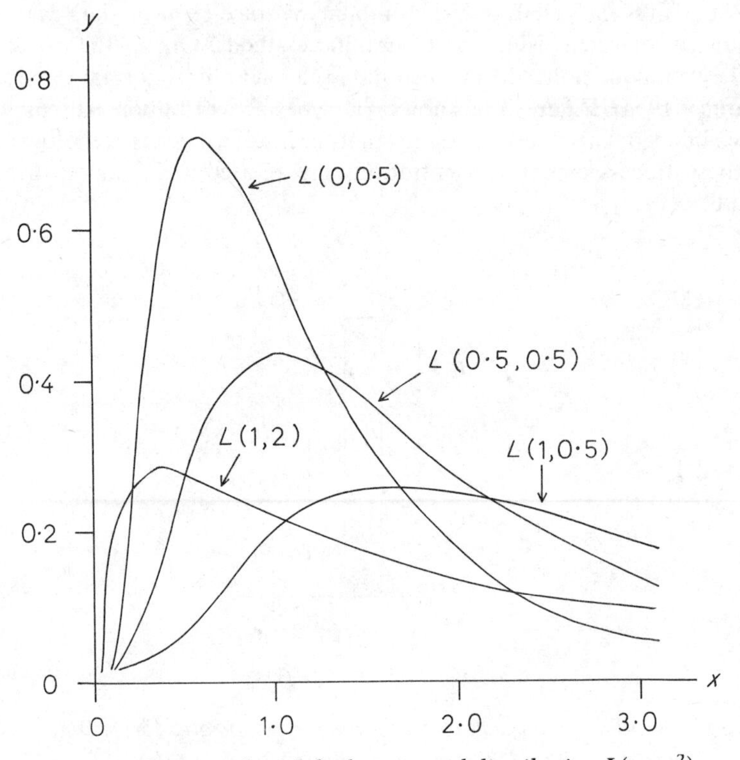

Figure 3.19 Examples of the log–normal distribution $L(\mu_n, \sigma_n^2)$.

The log-normal distribution is important in the earth sciences, because examples of it have arisen in the study and search for economically important trace elements (see Koch and Link, 1970, chapter 6) or in studying precipitation and flow discharge in hydrology (see Yevjevich, 1971, chapter 7).

We have already accepted that grain size in a sediment is log-normally distributed by using the ϕ scale (see figures 3.1 and 3.2)

$$\phi = -\log_2 \text{(particle diameter in millimetres)}$$

gives

$$y = \frac{1}{\sigma_\phi\sqrt{(2\pi)}}\exp\left[-\frac{1}{2}\left(\frac{\phi-\mu_\phi}{\sigma_\phi}\right)^2\right]$$

which follows the normal distribution $N(\mu_\phi, \sigma_\phi^2)$.

In general we can work directly with our set of measurements (x in equation 3.7) and employ the log-normal distribution, or we can transform our measurements to logarithms and use the normal distribution (equation 3.4) with these $\ln x$ values. Some people find the latter easier to work with, because probabilities can then be related to z values (table 3.6). In addition, many statistical tests based on the normal distribution can then be used.

The Poisson distribution

Suppose a radioactive element decays by emitting α-particles. The rate of decay is defined by the half-life of the element, but α-particles are not emitted regularly, like bullets from a machine-gun. Over a short time-period the number of α-particles emitted (events) varies randomly. If we noted the number of events in each time interval, the mean value would be that defined by the half-life of the element. Values for each time interval could vary quite markedly and the distribution of these values would follow the Poisson distribution

$$y = \frac{\mu^x}{x!} \exp(-\mu) \tag{3.8}$$

where x is the number of events (for example, number of α-particles emitted in the time interval)

μ is the population mean and variance.

This is a discrete probability distribution (as is the binomial) where x must be a non-negative integer. As μ becomes large the Poisson distribution tends to a normal distribution. Equation 3.8 is completely defined by one parameter, μ, which is both the mean and the variance of the population.

A number of assumptions are made in deriving the Poisson distribution.

1. The events occur independently—that is, the emission of an α-particle does not depend on previous or following emissions in any way.

2. The probability of an event occurring is proportional to the length of time since the last event—that is, the longer the element exists without emitting an α-particle the more likely it is to emit one.

3. Two events cannot occur simultaneously.

Figure 3.20 shows the form of the Poisson distribution.

Apart from measurements of radioactivity in geology there are only a few cases where the Poisson distribution can be applied. The distribution of the occurrence of major earthquakes in a given time interval seems to be Poissonian. Koch and Link (1970, chapter 6) show that contrary to expectations, the occurrence of meteorites over a given area does not follow a Poisson distribution. This is probably because meteorite finds only represent a fraction of the total falls, whereas few major earthquakes can go unnoticed!

Curl (1966) shows that the frequency of proper cave entrances follows a Poisson distribution.

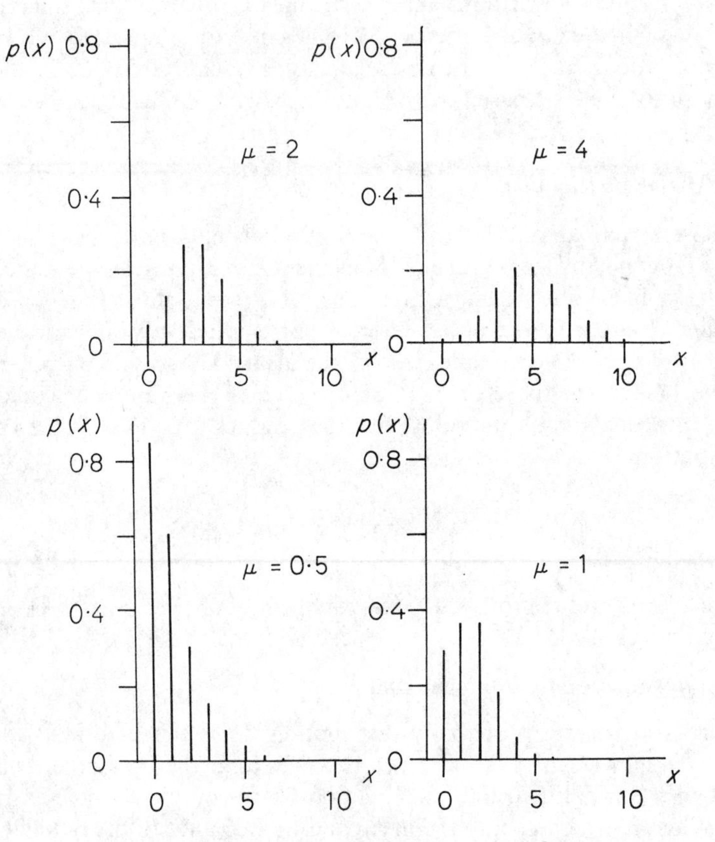

Figure 3.20 Examples of the Poisson distribution.

4 SAMPLING AND TESTS OF SIGNIFICANCE

We must now start to be more rigorous about the use of terminology in describing a distribution by following the accepted conventions. In considering the porosity of a sandstone body there was a population of all possible porosity measurements for that sandstone. Generally a population is a set of measurements (it is the measurements, not the objects) of a specified property of a group of objects. These measurements may be of attributes or of variables. Some people call this population a universe to distinguish it from a biological population—which is a sample. It may be an infinite population (every possible measurement of the length of this page) or a bounded population (measurement of the size of every quartz grain in a thin section). Many geological populations can be considered infinite though they are really finite. For example, measurements of the orientation of every felspar phenocryst in a lava, or every pebble on a beach can be thought of as an infinite population.

In studying the properties of any measurement (parameter) for a population, the aim is to define its distribution in that population. There may be innumerable reasons for doing this. Perhaps the orientation of felspar phenocrysts will tell us the plane in which a lava flowed, the grain size of a sand will tell us about the processes of deposition, or the shape of a valley will tell us about the mechanisms of erosion that produced that valley. We must always have an aim in view before we even begin making any measurements. The statistics we produce are only passive aids to our conclusions. There is reputed to be a sign in all computer rooms 'Rubbish in, rubbish out'. It applies equally to the use of statistics.

It is not economical, of time or money, to measure the grain size of every pebble on a beach. So we take a small sub-set of the population, a sample, which we choose to be representative of the population parameter in which we are interested. Ideally we design our sampling procedure, having identified the properties which interest us. Regretfully, the nature of earth science work, such as mapping rocks, soils or geochemical anomalies, which is governed by natural exposure, may make this difficult. This leaves a problem—for the usefulness of a statistical result depends on the care taken in designing the sampling procedures.

Having taken a sample we then use its properties (parameters) as estimators of the population parameters. There is also a convention applied to parameters. Those of a population are represented by Greek letters—so μ is a population mean and σ^2 is its variance. Sample parameters are represented by Roman letters—m (or $\bar{x}$) is a sample mean and its variance is s^2. 'Hats' can be used to represent sample estimates of a population parameter. The best sample estimates of μ and σ^2 can be written $\hat{\mu}$ and $\hat{\sigma}^2$ (but $\hat{\mu}$ will probably be the same as $\bar{x}$).

This use of a sample of objects to find properties of the parent population is called statistical inference. By the application of suitable statistics we try to infer some conceptual model for our observations. We may then use any non-statistical reasoning to extend these properties to a wider target population. For example, we may sample a granite batholith (the population) to define its average quartz content (sample estimate of the population mean). We may then wish to use this as the average quartz content of all south of England granites (the target population) in some geological context.

4.1 Sampling procedures

Two main types of sample are collected; the systematic and the random sample. With a systematic sample, information will be spread evenly over the sample space. Objects are chosen on a grid (hand specimens from a fully exposed intrusion or the point counting of a thin section) or a regular interval (shear strength of a deep-sea clay every metre down a core). This even distribution of data points is useful in some statistical procedures.

A random sample, though suggesting to some people an odd, thoughtlessly collected sample, is actually very carefully obtained. It is random because each object in the population has the same chance of being selected every time an object is taken. This sampling procedure is said to be unbiased.

Actually collecting a random sample can be difficult, simply to ensure that every object has an equal probability of being chosen. Schemes can range from picking objects out of a bag (which is shaken between selections) to throwing a geological hammer over your shoulder at a quarry face. Alternatively objects, or their locations, can be numbered and then chosen, by taking numbers from random-number tables. Sets of random numbers, usually computer generated, can be found in most statistical tables. Part of such a table is reproduced in table 4.1. A few simple examples should give a clear idea of how to use random-number tables. We are often told to use them, but the actual use is rarely explained.

The random numbers are generated so that every integer between zero and nine has an equal chance of being chosen for every entry. The numbers

Table 4.1 Part of a table of random numbers, taken from Murdoch and Barnes (1974, table 24) with permission.

03 47 43 73 86	36 96 47 36 61	46 98 63 71 62	33 26 16 80 45	60 11 14 10 95
97 74 24 67 62	42 81 14 57 20	42 53 32 37 32	27 07 36 07 51	24 51 79 89 73
16 76 62 27 66	56 50 26 71 07	32 90 79 78 53	13 55 38 58 59	88 97 54 14 10
12 56 85 99 26	96 96 68 27 31	05 03 72 93 15	57 12 10 14 21	88 26 49 81 76
55 59 56 35 64	38 54 82 46 22	31 62 43 09 90	06 18 44 32 53	23 83 01 30 30
16 22 77 94 39	49 54 43 54 82	17 37 93 23 78	87 35 20 96 43	84 26 34 91 64
84 42 17 53 31	57 24 55 06 88	77 04 74 47 67	21 76 33 50 25	83 92 12 06 76
63 01 63 78 59	16 95 55 67 19	98 10 50 71 75	12 86 73 58 07	44 39 52 38 79
33 21 12 34 29	78 64 56 07 82	52 42 07 44 38	15 51 00 13 42	99 66 02 79 54
57 60 86 32 44	09 47 27 96 54	49 17 46 09 62	90 52 84 77 27	08 02 73 43 28
18 18 07 92 46	44 17 16 58 09	79 83 86 19 62	06 76 50 03 10	55 23 64 05 05
26 62 38 97 75	84 16 07 44 99	83 11 46 32 24	20 14 85 88 45	10 93 72 88 71
23 42 40 64 74	82 97 77 77 81	07 45 32 14 08	32 98 94 07 72	93 85 79 10 75
52 36 28 19 95	50 92 26 11 97	00 56 76 31 38	80 22 02 53 53	86 60 42 04 53
37 85 94 35 12	83 39 50 08 30	42 34 07 96 88	54 42 06 87 98	35 85 29 48 39
70 29 17 12 13	40 33 20 38 26	13 89 51 03 74	17 76 37 13 04	07 74 21 19 30
56 62 18 37 35	96 83 50 87 75	97 12 25 93 47	70 33 24 03 54	97 77 46 44 80
99 49 57 22 77	88 42 95 45 72	16 64 36 16 00	04 43 18 66 79	94 77 24 21 90
16 08 15 04 72	33 27 14 34 09	45 59 34 68 49	12 72 07 34 45	99 27 72 95 14
31 16 93 32 43	50 27 89 87 19	20 15 37 00 49	52 85 66 60 44	38 68 88 11 80
68 34 30 13 70	55 74 30 77 40	44 22 78 84 26	04 33 46 09 52	68 07 97 06 57
74 57 25 65 76	59 29 97 68 60	71 91 38 67 54	13 58 18 24 76	15 54 55 95 52
27 42 37 86 53	48 55 90 65 72	96 57 69 36 10	96 46 92 42 45	97 60 49 04 91
00 39 68 29 61	66 37 32 20 30	77 84 57 03 29	10 45 65 04 26	11 04 96 67 24
29 94 98 94 24	68 49 69 10 82	53 75 91 93 30	34 25 20 57 27	40 48 73 51 92
16 90 82 66 59	83 62 64 11 12	67 19 00 71 74	60 47 21 29 68	02 02 37 03 31
11 27 94 75 06	06 09 19 74 66	02 94 37 34 02	76 70 90 30 86	38 45 94 30 38
35 24 10 16 20	33 32 51 26 38	79 78 45 04 91	16 92 53 56 16	02 75 50 95 98
38 23 16 86 38	42 38 97 01 50	87 75 66 81 41	40 01 74 91 62	48 51 84 08 32
31 96 25 91 47	96 44 33 49 13	34 86 82 53 91	00 52 43 48 85	27 55 26 89 62
66 67 40 67 14	64 05 71 95 86	11 05 65 09 68	76 83 20 37 90	57 16 00 11 66
14 90 84 45 11	75 73 88 05 90	52 27 41 14 86	22 98 12 22 08	07 52 74 95 80
68 05 51 18 00	33 96 02 75 19	07 60 62 93 55	59 33 82 43 90	49 37 38 44 59
20 46 78 73 90	97 51 40 14 02	04 02 33 31 08	39 54 16 49 36	47 95 93 13 30
64 19 58 97 79	15 06 15 93 20	01 90 10 75 06	40 78 78 89 62	02 67 74 17 33
05 26 93 70 60	22 35 85 15 13	92 03 51 59 77	59 56 78 06 83	52 91 05 70 74
07 97 10 88 23	09 98 42 99 64	61 71 62 99 15	06 51 29 16 93	58 05 77 09 51
68 71 86 85 85	54 87 66 47 54	73 32 08 11 12	44 95 92 63 16	29 56 24 29 48
26 99 61 65 53	58 37 78 80 70	42 10 50 67 42	32 17 55 85 74	94 44 67 16 94
14 65 52 68 75	87 59 36 22 41	26 78 63 06 55	13 08 27 01 50	15 29 39 39 43
17 53 77 58 71	71 41 61 50 72	12 41 94 96 26	44 95 27 36 99	02 96 74 30 83
90 26 59 21 19	23 52 23 33 12	96 93 02 18 39	07 02 18 36 07	25 99 32 70 23
41 23 52 55 99	31 04 49 69 96	10 47 48 45 88	13 41 43 89 20	97 17 14 49 17
60 20 50 81 69	31 99 73 68 68	35 81 33 03 76	24 30 12 48 60	18 99 10 72 34
91 25 38 05 90	94 58 28 41 36	45 37 59 03 09	90 35 57 29 12	82 62 54 65 60
34 50 57 74 37	98 80 33 00 91	09 77 93 19 82	74 94 80 04 04	45 07 31 66 49
85 22 04 39 43	73 81 53 94 79	33 62 46 86 28	08 31 54 46 31	53 94 12 38 47
09 79 13 77 48	73 82 97 22 21	05 03 27 24 83	72 89 44 05 60	35 80 39 94 88
88 75 80 18 14	22 95 75 42 49	39 32 82 22 49	02 48 07 70 37	16 04 61 67 87
90 96 23 70 00	39 00 03 06 90	55 85 78 38 36	94 37 30 69 32	90 89 00 76 33

are in column and row blocks purely for visual effect and we can dip into the table at any point.

Suppose we have ten brachiopods and want to measure the lengths of a sample of three of them—which three should we choose? To number the specimens 0 to 9 and then choose three random numbers would be an unbiased way of sampling the fossils. To get the random numbers let us start in column-block 3 and row-block 6 of table 4.1 (any starting place is acceptable). The first three numbers are 6, 7, 1—so we measure brachiopods 6, 7 and 1. If we have one hundred fossils and want to choose ten, then we can number the fossils 0 to 99, and starting at the beginning of table 4.1 we find 03, 47, 43, 73, 86, 36 ... etc., which gives the numbers of specimens on which we perform the measurements.

Using the tables is very easy if the size of a population is some power of ten. If we want to choose 5 out of 30 specimens, then the objects can be numbered 0 to 29, and choosing a place to begin reading our random numbers (say column-block 2, row-block 6 in table 4.1) we find 83, 62, 64, 11 and 12. Three of these random-numbers are outside the range of our specimen numbers. The easiest procedure is to divide all the selected numbers by 30 and use the remainder to choose specimens—this gives 23, 02, 04, 11 and 12.

Grid locations can also be chosen at random from these tables. Suppose we wish to sample a gabbro intrusion for geochemical analysis. We could map out a 100×100 grid over the area and number each row and column of the grid 0 to 99. It would be impracticable to collect and analyse the rocks at every grid intersection. Using the random-number tables to generate row and column numbers, we can select a random sample of specimens for analysis. Starting at the last block of table 4.1 we read 45, 07, 31, 66, 49, 85, 22, 04 ... etc., and collect samples from row 45, column 07; row 31, column 66; ... etc., of our mapped grid.

These procedures are quite simple and ensure that an unbiased sample is collected to define the population of interest. There are many variations which the worker can use. For example, one might not sample a circular intrusion on a square grid, but would use some form of circular mesh; or in sampling a delta lobe a fan-shaped mesh might be appropriate. A little thought and planning prior to leaping into the field can save a lot of time, effort and expense later.

Some form of pseudo-random sampling is often used because it is more convenient and still gives us a good representative sample. For example, in sampling a gravel we may pour a bag of it into a heap, 'cut' the heap in four, choose one quarter randomly and measure the size of all the pebbles in that sample. This type of pseudo-random sampling is often quicker and easier then random sampling. It is perfectly acceptable if the sampling rule is in no way related to the property being sampled. Obviously we do not take the twenty largest pebbles to get an average grain size for the gravel!

In systematic sampling we often use a random process to start off the sequence. For example the co-ordinates of the centre of a sampling grid may be chosen with random numbers.

Sampling procedure is most important because our tests and conclusions depend on this first step. Usually the sample is a very small fraction of the original population. For example, if we estimate the composition of a rock from the modal analysis of a thin section, we count a few thousand grains on a section that is 30 μm thick and about 5 square centimetres in area. This may be taken from a 1 dm^3 hand specimen, which is supposed to represent many cubic metres of rock. We make estimates about a population of thousands of millions of grains from counting a few thousand grains. Clearly our sample must be very carefully chosen.

Often the actual sampling procedure is not thought about carefully enough. At a conference on sampling methods (*Institute of Mining and Metallurgy*, London 1971) Professor J. C. Griffiths of Pennsylvania State University said, 'sampling is like religion—everyone is for it but no-one practises it.' Another participant related a case where grid sampling was used in exploration for diamonds. This sounds reasonable, except that the diamonds were expected to be located in the gravels in abandoned river channels —linear features whose approximate position was known!

Krumbein and Graybill (1965, p. 164) list the steps in sampling a population. The following points are paraphrased from them

1. Development of a set of concepts for the population of interest. Choice of variables to be measured and recognition of their sources of variation.

2. Translation of the set of concepts into a statistical model whose mathematical form includes all the known sources of variation within the population.

3. Selection of a sampling plan to suit the statistical model.

4. Decision about the number of samples to collect.

The last step often requires a preliminary sample to be taken. This allows the variance of the population to be assessed approximately, so that a sample size can be recommended to give a specific level of confidence to the results.

A good treatment of more complex sampling methods is given in Krumbein and Graybill (1965) and Griffiths (1967), and with particular reference to economic geology in Koch and Link (1970). Various authors contributed to the *Institute of Mining and Metallurgy* conference on sampling (*Trans. Inst. Mining Metall. Section B*, **80**, November 1971). Students should explore sampling problems themselves by using different schemes to estimate some population parameters. For example, try using various grid sizes on a rock surface, or devising different ways of estimating the grain size of a sack of gravel.

4.2 Some important sample properties

The distribution of sample means

Suppose we have a population consisting of only six specimens of a new species of ostracod, whose length we have measured as part of our biometrical summary of the species. These lengths were

$$2.5, \quad 3, \quad 3.5, \quad 3.5, \quad 4, \quad 4.5 \text{ mm}$$

which approximates a normal distribution N (3.5, 15/36).

If we take repeated random samples of two ostracods from this population we can demonstrate a very important rule of statistics. For a true random sample every object in the population must have an equal chance of being chosen every time an object is picked. This means we use sampling with replacement. Suppose we put one of our ostracods in each of the grids numbered 1 to 6 on a microscope slide. Then we choose a pair of measurements to use by throwing a pair of fair dice. Every ostracod has an equal chance of being chosen for each number of every sample. If the dice both land at 1, then we use the measurement for the ostracod in grid space 1 twice. This may seem strange, but it is only like drawing a sample from a bag, replacing it, shaking the bag and drawing it again—this is just sampling with replacement.

Taking samples of size two ($n = 2$) from a population of six ($N = 6$) there are 6^2 (N^n) possible samples, containing the following measurement pairs

2.5, 2.5; 2.5, 3; 2.5, 3.5; 2.5, 3.5; 2.5, 4; 2.5, 4.5; 4.5, 2.5;
4, 2.5;... 4, 4; 4, 4.5; 4.5, 4; 4.5, 4.5.

Note that a, b is a different sample from b, a, because the order of selection differs. If the list of possible samples is complete, we can calculate the mean value of ostracod length for each sample. Each mean value, $\bar{x}$, is a sample estimate of the population mean. What would the distribution of these mean values look like? It is plotted in figure 4.1 and looks suspiciously like a normal distribution (which it in fact is) with a mean of 3.5 mm. But this is the population mean, so for our case the mean of sample means is the population mean. We say the mean of sample means is an unbiased estimator of the population mean.

Perhaps the variance of this distribution of sample means also has some useful properties? If we calculate this variance of sample means around the population mean (which is the mean of the sample means) thus

$$\sigma_{\bar{x}}^2 = \sum_{i=1}^{36} (\bar{x}_i - \mu)^2/n$$

where $\bar{x}_i$ = the mean value of the ith sample, we find that its value is 15/72. This is half the population variance, for our sample size of two, that is

$$\sigma_{\bar{x}}^2 = \sigma^2/n \qquad (n = 2 \text{ here})$$

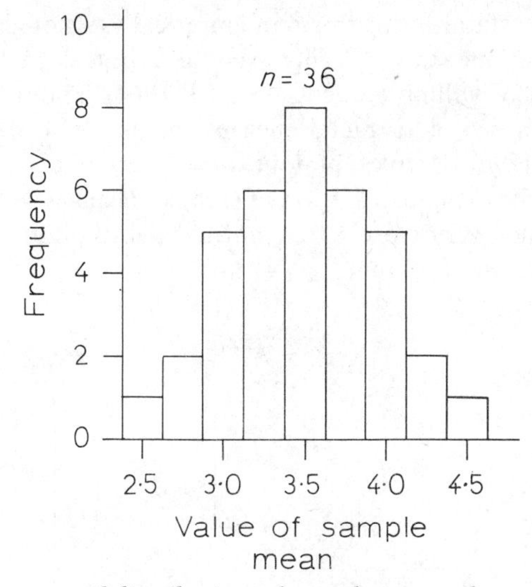

Figure 4.1 Histogram of distribution of sample means for two-object samples from the ostracod population.

These relationships could be checked empirically for other sample sizes and they can, of course, be proved mathematically. However we have deduced two very important properties, which can be applied to the sampling of attributes of variables as stated below.

If we take all possible samples of size n, from a population with mean μ and variance σ^2, these sample means will be distributed with mean $\mu_{\bar{x}}$ and variance $\sigma_{\bar{x}}^2$ where

$$\mu_{\bar{x}} = \mu \qquad \text{and} \qquad \sigma_{\bar{x}}^2 = \sigma^2/n \tag{4.1}$$

These properties do not only apply to populations which are normally distributed, but if the population is a normal distribution $N(\mu, \sigma^2)$, then the sample means will automatically be normally distributed $N(\mu, \sigma^2/n)$.

The central limit theorem

This also is an important statistical theorem, whose conclusions are invoked in making tests.

If we repeat our sampling experiment with a different population of ostracod measurements with a non-normal distribution, the rules about the mean and variance of sample means would still apply as defined above. In addition, we would find that as we made our sample size, n, larger the distribution of sample means approximates more and more to a normal distribution. This will occur for virtually all types of distributions and certainly for all those we are likely to encounter.

We can easily establish this theorem empirically—though we would have to calculate many means (N^n). This exercise is best done by computer, or by a large class of willing students (see Li, 1964). Figure 4.2A shows the distribution of a set of ostracod measurements ($N = 6$) from which all possible ($6^4 = 1296$) samples of four ($n = 4$) were chosen, by a simple computer program. The distribution of sample means is seen in figure 4.2B, which corresponds very closely to a normal distribution $N(3.0, 0.04)$, with a standard deviation of 0.2 mm.

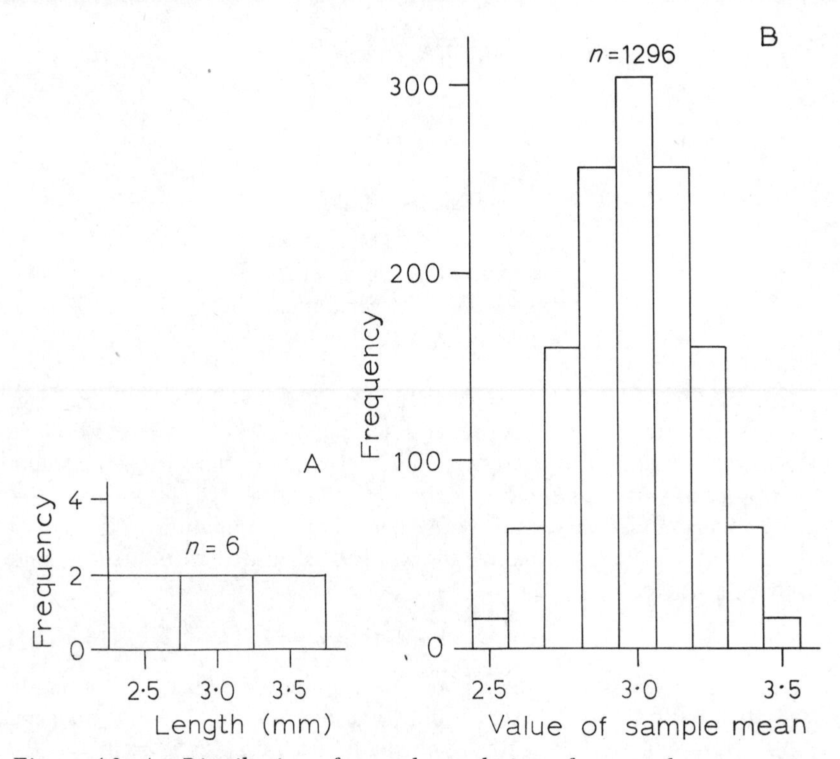

Figure 4.2 A—Distribution of second population of ostracod measurements; B—Distribution of sample means for all possible samples of size four ($n = 4$) from figure 4.2A.

This example is a case of the central limit theorem and says that whatever the form of the population distribution, the distribution of the means of all possible samples of size n from that population, tends to a normal distribution as n becomes infinite. Combining this with the previous conclusions, we can now approximate the distribution of sample means from any population, with mean μ and variance σ^2, to a normal distribution $N(\mu, \sigma^2/n)$. If the original population is actually normal then the sample means are normally distributed.

The sample variance

We have been relating our sample estimates to the population parameters which we knew. This is not commonly the case in practice, since we are likely to be calculating statistics to obtain estimates of the population parameters. So let us look at the variance of each sample about its own mean, which is known, rather than about the population mean, which will not normally be known. We can again use our normally distributed ostracod data $N(3.5, 15/36)$ from page 55. Will the average value of this individual sample variance for all N^n samples be the population variance? Using

$$S^2 = \frac{\sum_{i=1}^{n} (x_i - \bar{x})^2}{n} \qquad (n = 2)$$

with $\bar{x}$, rather than μ we get values of S^2

0; 1/16; 1/4; 1/4; 9/16; 1; 1; 9/16; ... 0; 1/16; 1/16; 1

when the sample pairs are arranged as before. These 36 values of S^2 have a mean value of 15/72, which is not the population variance. In fact it is a biased estimator of σ^2, tending to underestimate it. However, if we define an s^2

$$s^2 = \frac{\sum_{i=1}^{n} (x_i - \bar{x})^2}{n - 1} \tag{4.2}$$

this has a mean value of 15/36 which is the population variance. We shall always in future use this s^2 and call it the sample variance, because it is the best, unbiased, sample estimator of the population variance. (The efficiency of s^2 can be checked further using any real or artificial set of data from which samples of size n could be picked).

Equation (4.2) is the correct way of defining s, the sample variance, being the sum of squares of deviations of x about the sample mean $\bar{x}$. An easier, but equivalent computational form is usually used

$$s^2 = \frac{\left[\sum_{i=1}^{n} x_i^2 - \left(\sum_{i=1}^{n} x_i\right)^2 \Big/ n\right]}{n - 1} \tag{4.2a}$$

Equation (4.2a) is easier to compute on a calculator, since most modern calculators allow the $\sum x^2$ and the $\sum x$ to be accumulated simultaneously with only one pass through the data. The form of equation (4.2) requires the calculation of a mean value and then the summing of deviations about that mean. This needs two passes through the data set. It is easy to prove the equivalence of equations (4.2) and (4.2a).

The number $(n - 1)$ in equations (4.2) and (4.2a) is referred to as the degrees of freedom of the sample. Without recourse to a page of mathematics, degrees of freedom are difficult to define. Yet we need to know them for most statistical tests. An analogy can be made with the Phase Rule in chemistry, where the degrees of freedom are the unconstrained parameters (concentrations, temperature or pressure) in a system. Similarly in statistics our variance has only $(n - 1)$ unconstrained values, for once $(n - 1)$ values of x_i have been defined the nth value is also known since

$$\sum_{i=1}^{n} (x_i - \bar{x}) = \sum_{i=1}^{n} x_i - n\bar{x} = 0$$

because

$$\bar{x} = \sum_{i=1}^{n} x_i/n$$

In general the degrees of freedom are the number of observations minus the number of estimates made from them.

4.3 Student's *t*-distribution and an introduction to hypothesis testing

The ideas developed in this section are central to the use of statistical tests and they need to be read very thoughtfully, before moving on to subsequent chapters. Though hypothesis testing is introduced and explained through the use of Student's t-distribution, the following chapters show many other occasions where it is applied. In all the following sections reference to 'a sample' means a 'random sample'. This sample is assumed to be taken in such a way that every member of a population has an equal chance of being selected.

Student's t-distribution

William S. Gosset was a statistician who defined a distribution for sample means in the first decade of this century. He always wrote under the pseudonym of 'Student'. When R. A. Fisher, a very famous statistician modified this distribution and christened it 't' it became known as Student's t-distribution.

If we take all possible random samples of size n from a normal distribution $N(\mu, \sigma^2)$ and calculate

$$t = \frac{\bar{x} - \mu}{s/\sqrt{n}} \tag{4.3}$$

for each sample, we will get a distribution of t values. (s = sample standard deviation, with devisor $n - 1$.)

The statistic t has a distribution as shown in figure 4.3. (We could generate this from a small set of data, as an exercise similar to that in section 4.2.)

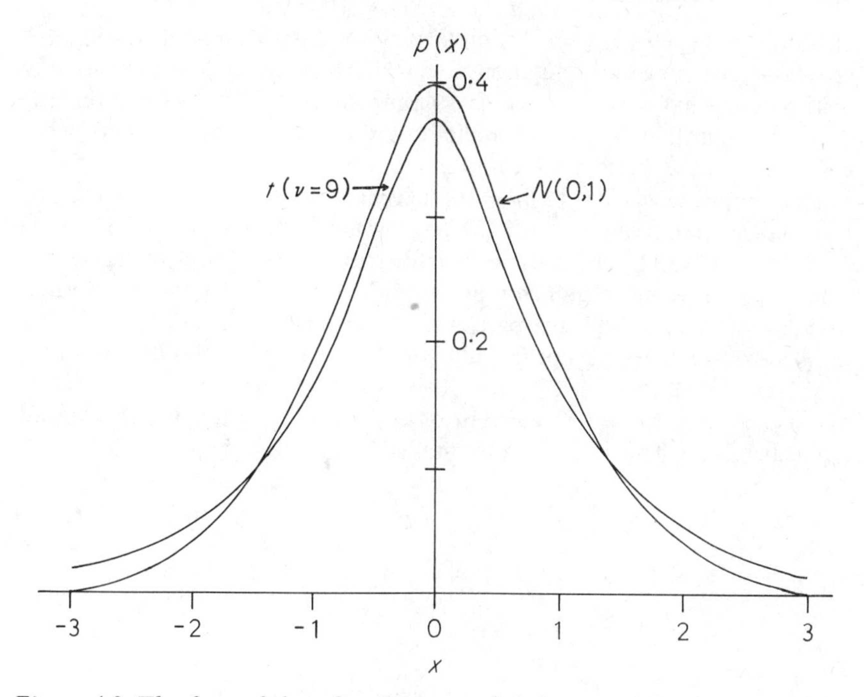

Figure 4.3 The form of the t-distribution with 9 degrees of freedom, compared with N(0, 1) the standardised normal distribution.

The form of the t-curve looks very like the standardised normal curve, which is also shown in figure 4.3. For values of n greater than 30, the t-distribution tends to a normal distribution. For smaller values of n the curves differ quite markedly in the tails of their distributions. The similarity of these curves is not surprising, since the formula for t looks very like the z we defined in section 3.3, except that we are using s in place of σ. t has $n - 1$ degrees of freedom because with a given n objects we can calculate s and $\bar{x}$ freely, but must estimate μ.

This t-distribution is very useful, because we can test populations whose σ we do not know and can predict likely values of μ, the population mean. Table 4.2 gives a sample of volcanic gas compositions for a recent eruption

Table 4.2 Sulphur content (in per cent number of atoms) of gases collected from the 1970 Mount Etna volcanic eruption. Data provided by T. Huntingdon.

13.3	11.4	9.7	9.3
5.9	7.0	11.6	12.8
4.7	9.9	10.5	5.7
8.0	9.8	6.7	11.8

of Mount Etna. This is a small sample of all the gases emanating during this eruption. Assuming the population from which this sample comes is normally distributed, what was the average sulphur content of the gases from the 1970 eruption (that is, can we predict μ, the population mean)? With what confidence can we state a value and how is confidence quantified?

The sample consists of 16 ($n = 16$) determinations of sulphur content for gases taken from fumeroles in 1970. Any t we calculate will have 15 ($n - 1$) degrees of freedom. The form of this distribution is shown in figure 4.4. Following the same arguments as we did in section 3.3 for the normal distribution, we can find that part of the t-curve which contains 95 per cent of its area. We can look up the value of t in statistical tables (for example, Murdoch and Barnes, 1974, table 7). If we take this area symmetrically about the mean, we get 2.5 per cent of the area left outside at either tail of the distribution. This is called a two-tailed value and is shown in figure 4.4.

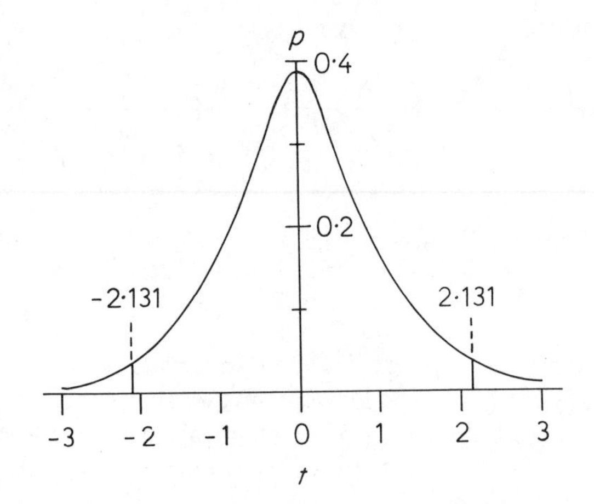

Figure 4.4 A two-tailed 5 per cent region for t with 15 degrees of freedom.

We actually look up a '2.5 per cent value of t, with 15 degrees of freedom' in the tables—that is the value of t beyond which 2.5 per cent of the area of the curve lies at each tail. This t-value is 2.131.

When random samples of size 16 are taken from a normal population $N(\mu, \sigma^2)$ only 5 per cent of all samples will have such extreme values of $\bar{x}$ and s that they give t values outside the range $+2.131$ to -2.131. Furthermore, 2.5 per cent of t values would be greater than $+2.131$ and 2.5 per cent of t values would be less than -2.131. The probability of t lying between $+2.131$ and -2.131 must be 0.95, since this range includes 95 per cent of the possible values that t can take, and the probability of t being outside that range (that is, lying in a critical region) is 0.05 (being 0.025 for each tail). In mathematical form we write a two-tailed t

$$t_{(\alpha/2;\, v)} = t_{(0.05/2;\, n-1)} = t_{(0.025;\, 15)} = 2.131$$

where v (the Greek letter nu) is used for degrees of freedom $= n - 1$ and α is the probability of obtaining a value greater than $|t|$ from a normal distribution $N(\mu, \sigma^2)$.

Returning to the gas analyses we would expect that if we took a sample of 16 measurements from a population with mean μ, t would lie between ± 2.131 for 95 per cent of such possible samples. We defined t from equation 4.3

$$t_{(\alpha/2;\, v)} = \frac{\bar{x} - \mu}{s/\sqrt{n}}$$

where $v = n - 1$

$$-2.131 \leqslant \frac{\bar{x} - \mu}{s/\sqrt{16}} \leqslant +2.131 \tag{4.4}$$

rearranging t in equation (4.4) and substituting the correct t value we get

$$\bar{x} + 2.131s/\sqrt{16} \leqslant \mu \leqslant \bar{x} + 2.131s/\sqrt{16} \tag{4.5}$$

which says that we are 95 per cent certain that the population mean (μ) we are about to estimate from our sample of sixteen gases will lie in the range $\bar{x} \pm 2.131s/\sqrt{16}$. This is deduced before any sample of measurements is even collected. We now take our sample, determine the gas contents, and calculate $\bar{x}$ and s.

For the 1970 gas data

$$\bar{x} = 148.1/16 = 9.26$$

$$s = (1476.65 - 1370.85)^{\frac{1}{2}}/15 = (7.053)^{\frac{1}{2}} = 2.66$$

$$7.84 \leqslant \mu \leqslant 10.68$$

We can now say that there is a probability of 0.95 that the percentage of sulphur atoms in the total population of 1970 Etna volcanic gases lies between 7.84 and 10.68. We have accepted a small probability (0.05) that we have used a very extreme sample to estimate μ, such that μ is really outside the specified range, but we have quantified this risk. Formally we refer to a 95 per cent confidence interval (CI_{95}) on the population mean

$$CI_{95}\colon\ 7.84 \leqslant \mu \leqslant 10.68$$

Defining this confidence interval is a useful way of characterising a population from a sample. For example, CI can be used as an estimate of biological variation or of precision of an analysis (see section 4.6).

We were considering the likelihood of occurrence of a range of t-values distributed symmetrically about the mean (a two-tailed t-test giving a

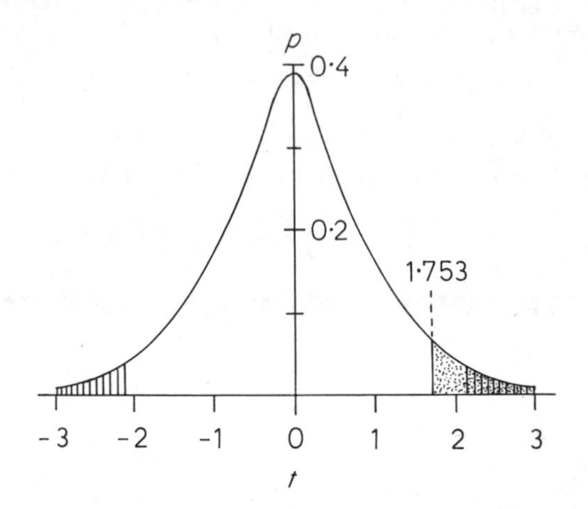

Figure 4.5 A one-tailed 5 per cent critical region for t with 15 degrees of freedom. The two-tail critical region of figure 4.4 is shaded.

central confidence interval). Equally we might be interested in discovering the likelihood of our sample representing μ greater than a given value (a one-tailed t-test giving a non-central confidence interval). Figure 4.5 shows the appropriate t-value for such a decision. Using our t-tables we find that 95 per cent of the area under the t-curve (for 15 degrees of freedom) lies below a t-value of 1.753 (that is, between $+1.753$ and $-\infty$). We write this one-tailed t-value

$$t_{(\alpha;\nu)} = t_{(0.05;n-1)} = t_{(0.05;15)} = 1.753$$

setting up our inequalities as before

$$-\infty \leqslant \frac{\bar{x} - \mu}{s/\sqrt{16}} \leqslant 1.753 \tag{4.6}$$

which rearranges to

$$\bar{x} - 1.7533\, s/\sqrt{16} \leqslant \mu \leqslant \infty \tag{4.7}$$

using the 1970 gas analyses this gives

$$8.09 \leqslant \mu \leqslant \infty$$

This means that we can say the probability of μ being greater than or equal to 8.09 is 0.95. There are only 5 out of 100 chances of us getting our sample, with its $\bar{x}$ and s value, from a population whose mean is outside this range. Note that setting an upper limit on $t(-\infty \leqslant t \leqslant 1.753)$ defines a lower confidence interval on the mean μ (called a greatest lower bound—GLB) written

$$\text{GLB}_{95} = 8.09; \qquad 8.09 \leqslant \mu \leqslant \infty$$

With a little thought we should expect this to be the case. If our value of $\bar{x}$ is an extreme value compared to the population mean, then there are two alternatives: either μ is much larger than $\bar{x}$, which would give a negative t value, or μ is small and $\bar{x}$ is an extremely large estimate (giving a large positive t-value). To be precise, a t-value greater than 1.753 will occur 5 out of a 100 times (or less) from a μ-value of 8.09 (or smaller). This is why the inequalities 4.6 and 4.7 are reversed. This also means that in the central confidence limit the positive t-value defines the lower limit on μ and the negative t-value defines the upper limit on μ (see inequalities 4.4 and 4.5).

We could also set a least upper bound (LUB) on μ by considering t-values greater than -1.753

$$\text{LUB}_{95} = 10.42; \qquad -\infty \leqslant \mu \leqslant 10.42$$

which says we can be 95 per cent certain that our $\bar{x}$ and s represent a population mean μ equal to or smaller than 10.42.

The choice of a central or non-central confidence interval for a t-value depends entirely on our aim—that is, on what we wish to prove by the statistical procedure. Use of a t-test assumes that our sample does come from a normal distribution $N(\mu, \sigma^2)$ since this is the population from which a t-distribution is generated. However, a t-test can still be used if our population is non-normal, but it will be less powerful (see below for discussion of power). Often we can only assume that our sample comes from a normal distribution, yet we should be aware of our inherent assumption.

Testing hypotheses—(comparison of two means)

Table 4.3 gives some gas analyses for Mount Etna's eruptions in both 1970 and 1971. Vulcanologists want to know if there is a significant difference between the composition of the gases in the two eruptions. A t-test can also be used to answer this question, which is but one example of a commonly posed question (for example: Is there a significant drift between these two sets of gravimeter readings taken at a given location? Is there a significant difference in the sensitivity of these two muds?).

Suppose the two samples have means $\bar{x}_1$ and $\bar{x}_2$ and are of size n_1 and n_2

$$\mathbf{t} = \frac{\bar{x}_1 - \bar{x}_2}{S\sqrt{(1/n_1 + 1/n_2)}} \tag{4.8}$$

where

$$S = \left\{\frac{n_1 s_1^2 + n_2 s_2^2}{n_1 + n_2 - 2}\right\}^{\frac{1}{2}}$$

and is called the pooled sample variance, from samples with variance s_1^2 and s_2^2.

Table 4.3 Hydrogen content (in per cent number of atoms) of gases collected from the 1970 and 1971 Mount Etna volcanic eruptions. Data provided by T. Huntingdon.

1970		1971	
35.8	38.5	42.0	45.0
45.5	36.0	57.0	44.6
35.5	40.5	42.0	48.5
32.0	35.5	54.5	63.0
50.0	45.5	35.0	55.0
39.0	37.0	52.5	40.0
37.0	36.0	43.5	37.5
47.0	53.0	48.0	53.7

The statistic **t** in equation 4.8 follows the t distribution with $n_1 + n_2 - 2$ (that is, $n_1 - 1 + n_2 - 1$) degrees of freedom (d.f.). Repeated sampling of two normal populations (of equal variance) and calculation of **t** in equation 4.8 would give a distribution like figure 4.4.

To ask the question: Is there a significant difference between the mean values of the two populations from which our samples are drawn? we set up a hypothesis, called H_0, a null hypothesis to compare the two means. We ask how probable is it that our two samples, with their $\bar{x}$ and s values which are estimates of the population parameters, represent identical populations? This question is written

$$H_0(\mu_1 = \mu_2 | \sigma_1^2 = \sigma_2^2)$$

This null hypothesis states that the two samples come from two populations whose means μ_1 and μ_2 are equal. We are also assuming that the variance of the two populations is the same. (Assumptions are written to the right of the vertical line within the brackets.) This is a required condition since it was assumed in generating the possible distribution of t-values. If H_0 is true then the samples come from identical populations $N(\mu, \sigma^2)$ and can be regarded as two samples of the same population.

The purpose of the statistical test is to find out with what confidence we can accept or reject our hypothesis. If we can reject H_0, then μ_1 does not equal μ_2. This alternative hypothesis, H_a is usually the decision that interests us. Hence, some people refer to 'accepting H_0' as 'a failure to reject it'—our statements become even more convoluted!

The full test is now written

$$H_0(\mu_1 = \mu_2 | \sigma_1^2 = \sigma_2^2)$$

versus

$$H_a(\mu_1 \neq \mu_2)$$

For our two samples of gases (table 4.3)

$$\bar{x}_1 = 40.24 \qquad \bar{x}_2 = 47.61$$

$$s_1^2 = 36.8825 \qquad s_2^2 = 60.2599$$

$$n_1 = 16 \qquad n_2 = 16$$

giving **t** = 8.19, with 30 d.f.

For a direct comparison of means, where we do not wish to specify one μ as larger or smaller than the other, but merely as being different, we use a two-tail t-value. Now $t_{(0.05/2;30)} = 2.042$ from the tables, which is smaller than our value. This puts our **t**-value in the critical region and we can say that there are less than 5 in 100 chances of these two samples having the same population mean. So we reject the hypothesis H_0 at a 5 per cent significance level. Actually our value of **t** is greater than $t_{(0.0001/2;30)}$ which means there is less than one in ten thousand chances of our samples coming from the same population.

The power of a test

If our two samples discussed above were very extreme examples from two identical (that is, with the same mean) populations, then we would be making an error in rejecting H_0. This risk has been defined as having a probability of less than 0.05. This error is the significance level of the test, and is referred to as a type I error. It occurs when we conclude the population means are different when they are in fact the same—we are rejecting a true hypothesis H_0.

There is a second error, a type II error, that we could make by concluding that the population means are the same when they are different—we are accepting a false hypothesis H_0. Tabulating the conclusions of our test shows four possible results

Hypothesis H_0	Our test *accepts* H_0	Our test *rejects* H_0
Is really *true*	Correct conclusion	Type I error
Is really *false*	Type II error	Correct conclusion

expressed in terms of our test

$H_0(\mu_1 = \mu_2 \mid \sigma_1^2 = \sigma_2^2)$	Our test says $\mu_1 = \mu_2$	Our test says $\mu_1 \neq \mu_2$
Is really *true*	Correct conclusion	Type I error (α probability)
Is really *false*	Type II error (β probability)	Correct conclusion

But what is the probability, β, of a type II error? Often this is not known or even considered by the users of a statistical test. The power of a test is defined as $1 - \beta$. Since we are usually interested in establishing that H_0 is false, we should aim to minimise β, for a given α confidence level. Note that we can only make either a type I error or a type II error. The former can only arise if H_0 is true, the latter only if H_0 is false.

To understand how a type II error arises let us follow the logic of Li (1964, chapter 6) and consider another constituent of the Etna volcanic gases. Suppose we believe that the gases of the 1971 eruptions have a mean composition of 10 per cent carbon atoms—$\mu = 10$, with a variance (σ^2) of 4 per cent: that is the population is $N(10, 4)$. If we take samples of eight gas measurements ($n = 8$) then the mean values of all such possible samples should be distributed normally with mean $10(=\mu)$, and variance $4/8(=\sigma^2/n)$, giving $N(10, 4/8)$. The hypothesis H_0 is $H_0(\mu = 10|\sigma^2 = 4)$, versus $H_a(\mu \neq 10)$ for the population of gases. Figure 4.6 shows the expected distribution of sampling means $N(10, 4/8)$. Using our standard normal tables (table 3.6, pp. 34–5) we know that 95 per cent of all possible samples of size n taken from the population $N(\mu, \sigma^2)$ should have a mean value of $\bar{x}$ such that

$$\bar{x} \leqslant \mu \pm 1.96\sqrt{(\sigma^2/n)} \leqslant 10 \pm 1.96(4/8)^{\frac{1}{2}}$$

$$\bar{x} \leqslant 10 \pm 1.39$$

These critical values of the sampling mean are marked in figure 4.6. If the mean carbon content of our sample of eight gases lies beyond 11.39 or is less than 8.61 then we can say that there are less than 5 in 100 chances of it coming from a population with $\mu = 10$ (type I error is $\alpha = 0.05$).

Suppose that the sample mean was found to be eleven, then we would accept H_0 and say that at a 5 per cent confidence level we believe the population mean is $\mu = 10$. Now we could be making a type II error, because let us assume that the real population mean is 12 per cent carbon atoms. Our distribution $N(10, 4/8)$, which is curve A in figure 4.6, is therefore a hypothetical (and false) distribution. The true distribution is curve B. Any sample mean actually coming from distribution B, but having a mean value less than the defined critical point (11.39) will appear to come from curve A and we could only accept H_0. By accepting H_0 we would then be committing a type II error. The stippled area in figure 4.6 is β, the probability of a type II error. We can calculate this area of overlap in terms of its number of standard deviations to the left of the real mean ($\mu = 12$)

$$11.39 = \mu - d\sqrt{(\sigma^2/n)}$$

$$d = (12 - 11.39)/0.707$$

$$= 0.86 \text{ units of standard deviations to the left of the real mean.}$$

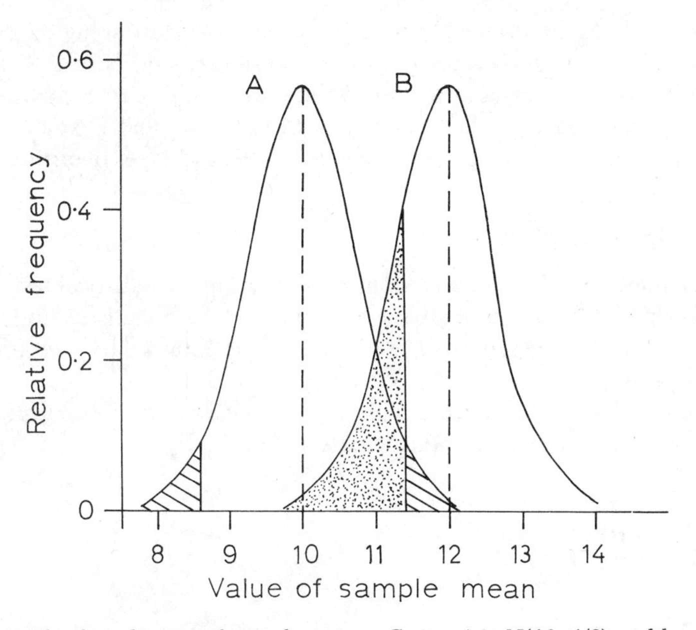

Figure 4.6 The distribution of sample means. Curve A is N(10, 4/8) and has the two-tail critical regions shaded. The area of type II error overlap of curve B, which is N(12, 4/8) is stippled.

So the stippled part of curve B contains the area under a normal curve beyond 0.86 standard deviations, which is 0.195 (from table 3.6, pp. 34–5).

Therefore β (probability of a type II error) $= 0.195$ and the power of our test $= 1 - \beta = 0.805$ or 80.5 per cent. From figure 4.6 we can make a number of suggestions about how to improve the power. If we increased our sample size, then the standard deviation of the distribution of sample means would be smaller (σ^2/n would decrease as n increases). This would reduce the overlap in the tails of curves A and B, because both curves would be narrower. For example if $n = 20$, the critical point on curve A would be at 10.875, giving $\beta = 0.006$ and a power of 99.4 per cent.

Reducing our α level of significance also decreases the area of overlap of the curves. For example if $\alpha = 0.10$, then the critical point on curve A is at 11.16, which gives $\beta = 11.7$, and a power of 88.3 per cent. Conversely increasing the α level of significance (that is reducing the chance of a type I error) say to $\alpha = 0.01$, increases the probability of a type II error. As the critical point of curve A moves to the right the type II area of overlap will become larger.

The type I and type II errors always work in opposition in this manner and we try to reach a reasonable compromise. This is usually found with $\alpha = 0.05$.

Statisticians naturally expend a lot of effort in establishing the power of a test. This is difficult because it depends on so many properties of the sample and its present distribution. Some tables give power curves for certain tests (for example, Pearson and Hartley, 1969, table 10) and Dixon and Massey (1957, chapter 14) give a lucid discussion of the two types of error and power.

4.4 The *F*-distribution

In chapter 6 we shall look at an important group of statistical tests based on a technique called the analysis of variance. These statistics follow the *F*-distribution, which is derived here because it is obtained by sampling normal populations.

If we take an independent sample from each of two normal populations with the same variance, and calculate a statistic

$$F = \frac{s_1^2}{s_2^2} \tag{4.9}$$

where s_1^2 is the variance of the sample from the first normal population

and s_2^2 is the variance of the sample from the second normal population.

If the samples are of size n_1 and n_2 the degrees of freedom of s_1 are $(n_1 - 1)$ and of s_2 are $(n_2 - 1)$.

Imagine one could take every possible sample of size n_1 from the first population and calculate s_1^2, and similarly calculate s_2^2 for the second population. From these s^2 values all possible F-values, as defined in equation 4.9 are calculated. This would produce an F-distribution with $n_1 - 1$ and $n_2 - 1$ degrees of freedom—called $F_{(\nu_1, \nu_2)}$ where $\nu_1 = n_1 - 1$ and $\nu_2 = n_2 - 1$. Both degrees of freedom must be defined since the range of s values will relate to the size of sample taken. Consequently an infinite family of F-curves exists. Figure 4.7 shows the form of the F-distribution for two pairs of degrees of freedom. The mean value of these distributions tends to unity as the number of degrees of freedom becomes large.

Provided that our samples are independent, repeatedly taking pairs of samples from a single normal distribution $N(\mu, \sigma^2)$ will also yield an F-distribution. Independent samples are those in which the choice of the members of one sample in no way governs or restricts the choice of members of the other sample.

Critical regions, which are one-tailed, can be defined for the F-distribution. Using tables (for example, Murdoch and Barnes, 1974, table 9) the F-value for $\alpha = 0.05$, with $\nu_1 = 4$ and $\nu_2 = 10$ is

$$F_{(0.05;4,10)} = 3.48$$

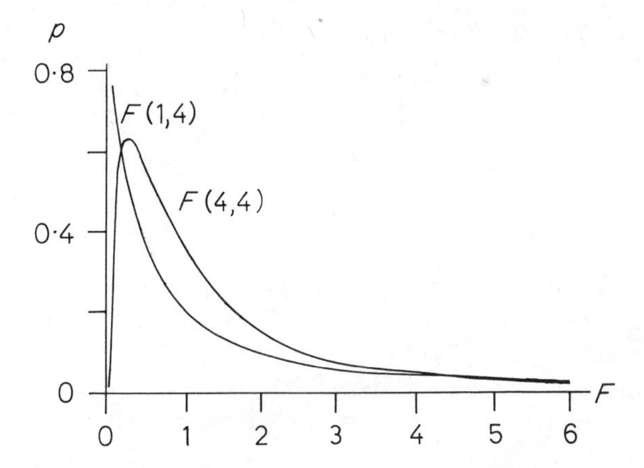

Figure 4.7 Examples of the F-distribution for various degrees of freedom.

which is marked in figure 4.8. Arguing exactly as for other distributions, 5 per cent of the area under the F-curve lies beyond an F-value of 3.48 (for F with 4 and 10 degrees of freedom).

Two properties of the F-distribution which can be derived are worth noting

(a) $F_{(1-\alpha;\, \nu_1,\, \nu_2)} = 1/F_{(\alpha;\, \nu_2,\, \nu_1)}$

For example

$$F_{(0.95;4,10)} = 1/F_{(0.05;10,4)} = 1/5.96 = 0.167 \qquad (4.10)$$

Tables normally only give the upper critical regions, because those are

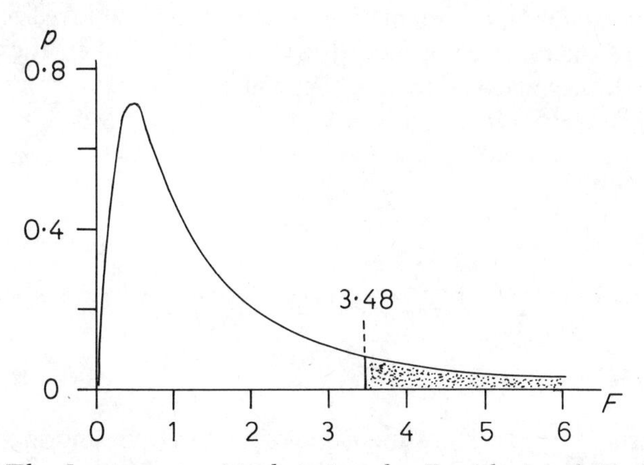

Figure 4.8 The 5 per cent critical region for F with 4 and 10 degrees of freedom.

usually used, but equation 4.10 does allow quick calculations of any lower-tail values.

(b) $F_{(\alpha;1,\nu_2)} = [t_{(\alpha/2;\nu_2)}]^2$

$F_{(0.05;1,10)} = 4.96 = (2.228)^2$

$t_{(0.025;10)} = 2.228$

F-tests are the staple diet of many experimental scientists. Their use in partitioning and analysing variance will be introduced in chapter 6, but we can already use an F-test in conjunction with the t-test. In comparing two sample means by a t-test it is assumed that the variance of the two populations represented by the samples is equal (page 62). An F-test will tell us if this assumption is in fact met and should be performed before the t-test.

Using the data for volcanic gases from table 4.3 we erect a hypothesis

$$H_0(\sigma_1^2 = \sigma_2^2 = \sigma^2) \text{ versus } H_a(\sigma_1^2 \neq \sigma_i^2)$$

that the variance of thw two samples is the same. From equation 4.9

$$F = s_1^2/s_2^2 = 60.3/36.88$$

$$F = 1.63 \text{ with 15 and 15 degrees of freedom}$$

Now $F_{(0.05;15,15)} = 2.40$, so we must accept H_0 and confidently (95 per cent) go on and use a t-test to compare the sample estimates of the population means.

4·5 The χ^2-distribution

Repeated sampling of a normal distribution is also employed to obtain the χ^2 (Greek chi and called chi-square) distribution. The mathematical derivation is given in Miller and Kahn (1962, appendix D), but a number of different forms of χ^2 are given in different statistical texts. Strictly it should be related to the standard normal deviate z defined in section 3.3

$$z = \frac{x - \mu}{\sigma}$$

If we take a sample of size n from $N(\mu, \sigma^2)$ and calculate z^2 for each object

$$\sum_{i=1}^{n} z_i^2 = \sum_{i=1}^{n} (x_i - \mu)^2/\sigma^2$$

Calculating $\sum z^2$ for all possible samples of size n from a normal population gives a distribution of $\sum z^2$ values which follows the form of a χ^2 distribution, as shown in figure 4.9.

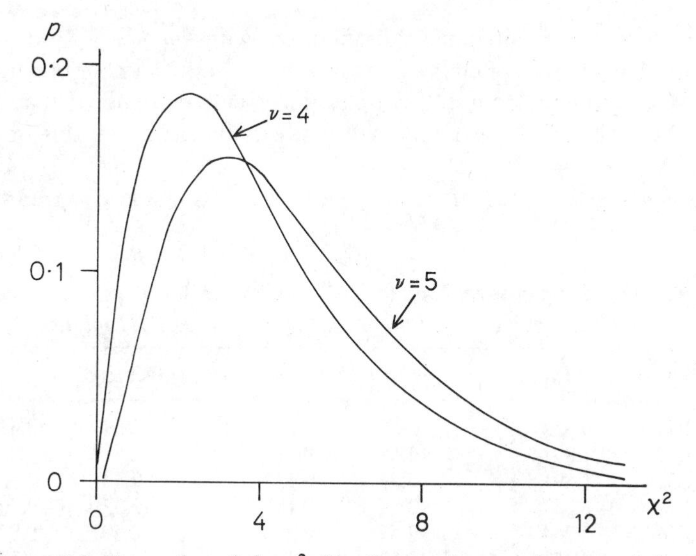

Figure 4.9 Examples of the χ^2-distribution for two degrees of freedom.

Goodness-of-fit

In addition a χ^2 distribution is also obtained from calculating all values of

$$F(x) = \frac{\text{(sum of squares)}}{\sigma^2}$$

for all possible samples of size n from $N(\mu, \sigma^2)$, where sum of squares is short for sum of squares of deviation about the mean

$$\text{Sum of squares (SS)} = \sum_{i=1}^{n} (x_i - \bar{x})^2$$

note: $s^2 = \text{SS}/(n-1) = \text{SS}/v$

$F(x)$ follows a chi-square distribution with $n - 1$ degrees of freedom (in defining $(n - 1)$ of the squared deviations from the sample mean the nth one is constrained since $\sum (x_i - \bar{x}) = 0$—see section 3.3).

The chi-square distribution is very important because it can be used in non-parametric (see chapter 7) as well as parametric tests. In particular it is often used to test the goodness-of-fit of a set of data to a theoretical distribution. A statistic X^2 is defined

$$X^2 = \frac{\text{(Observed value in the } i\text{th class} - \text{Expected value in the } i\text{th class)}^2}{\text{Expected value in the } i\text{th class}}$$

referred to as

$$X^2 = \frac{(\text{Observed} - \text{Expected})^2}{\text{Expected}} \quad \text{or} \quad \frac{(O - E)^2}{E} \tag{4.11}$$

which follows the chi-square distribution with $(m - k - 1)$ degrees of freedom. The number of classes used is m, and k is the number of parameters we have to estimate from our samples (for example, μ and σ^2 may be estimated by $\bar{x}$ and s^2). An example will best explain the use of this important test.

Table 4.4 gives a set of density measurements made on granite hand

Table 4.4 Density measurements (g/cm³) made on a set of granite hand specimens from a batholith. Data provided by D. T. Hopkins.

Density	z	Density	z	Density	z
2.59	0.332	2.59	0.332	2.59	0.332
2.62	1.891	2.58	−0.187	2.61	1.371
2.62	1.891	2.59	0.332	2.60	0.852
2.58	−0.187	2.58	−0.187	2.57	−0.706
2.54	−2.264	2.57	−0.706	2.60	0.852
2.60	0.852	2.57	−0.706	2.59	0.332
2.58	−0.187	2.59	0.332	2.56	−1.226
2.59	0.332	2.58	−0.187	2.58	−0.187
2.58	−0.187	2.58	−0.187	2.59	0.332
2.61	1.371	2.61	1.371	2.58	−0.187
2.60	0.852	2.58	−0.187	2.60	0.852
2.57	−0.706	2.59	0.322	2.54	−2.264
2.56	−1.226	2.57	−0.706	2.53	−2.784
2.58	−0.187	2.59	0.332	2.55	−1.745
2.58	−0.187	2.59	0.332	2.58	−0.187
2.57	−0.706	2.58	−0.187	2.59	0.332
2.59	0.332	2.62	1.891		

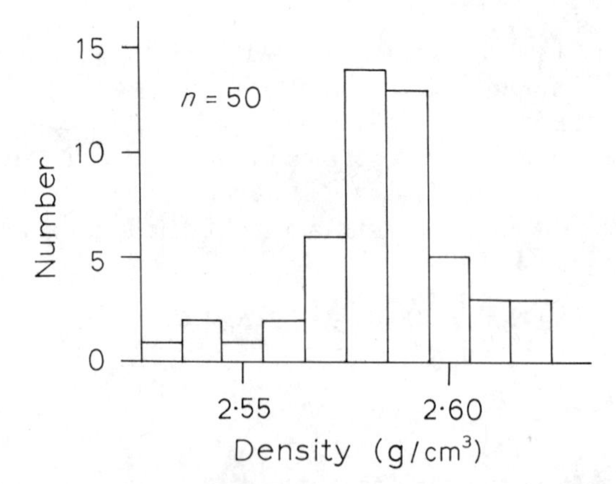

Figure 4.10 Histogram of granite density values for the data from table 4.4.

specimens collected from a batholith. Are the figures normally distributed? The procedure is to calculate the mean and variance of the data and to use these figures as sample estimates of μ and σ^2. The calculated X^2 value then tells us how likely the given distribution of values would be to occur in a sample from a normal distribution. We therefore set up a null hypothesis H_0 that the sample is from a normal distribution.

Figure 4.10 is a histogram of the density values, using the classes of table 4.5

$$\sum_{i=1}^{n} x_i = 129.18$$

$$\bar{x} = \sum x/n = 129.18/50 \qquad (n = 50)$$

$$\sum x^2 = 333.7676$$

$$s^2 = \frac{\sum x^2 - (\sum x)^2/n}{(n-1)} \qquad \text{(from equation 4.2a)}$$

$$= (333.7676 - 333.7494)/49$$

$$s^2 = 0.0003706$$

$$s = 0.01925$$

$$\bar{x} = 2.5836$$

Table 4.5 Frequency of occurrence of granite density values calculated from table 4.4.

Density range	Frequency	Relative frequency	Cumulative frequency
2.525–2.535	1	0.02	1
2.535–2.545	2	0.04	3
2.545–2.555	1	0.02	4
2.555–2.565	2	0.04	6
2.565–2.575	6	0.12	12
2.575–2.585	14	0.28	26
2.585–2.595	13	0.26	39
2.595–2.605	5	0.10	44
2.605–2.615	3	0.06	47
2.615–2.625	3	0.06	50

We compare our distribution to the standard normal distribution, so all density measurements need to be standardised, using

$$z = \frac{x - \bar{x}}{s} = \frac{x - 2.5836}{0.01925}$$

which expresses the deviation of each x value about the mean in units of standard deviation. These z-values are also given in table 4.4. We can then form a new histogram in terms of z-values with the following frequencies

z	Frequency		Class
<-2	3	6	<-1
-2 to -1	3		
-1 to 0	20	20	-1 to 0
0 to 1	18	18	0 to 1
1 to 2	6	6	>1
>2	0		
	50		

In table 4.4 the z-values for every density have been calculated. We can make our calculations shorter by turning the class intervals of table 4.5 into z-values, thus converting the complete histogram in one move.

To use the X^2-test we really need to have at least a frequency of five in each class. So classes are combined in the last column to give this situation. (Theory shows that when a number of classes are present containing less than five objects the X^2 calculated below starts to depart from the χ^2 distribution—this would introduce errors into our conclusions, that is, reduce the power of the test.) We obtain four classes of z: < -1, -1 to 0, 0 to $+1$ and $> +1$. Using our tables of areas under the normal curve (table 3.6) we find that the probability of a z-value lying beyond $+1$ is 0.1587, so a z-value between $+1$ and zero has a probability of $(0.5 - 0.1587)$, that is 0.3413, of occurring. If we take 50 z-values we expect 0.1587×50 z-values to be greater than $+1$. Thus we build up this table

z-class	Observed frequency	Probability of getting a z-value in this range	Expected frequency
<-1	6	0.1587	7.93
-1 to 0	20	0.3413	17.07
0 to $+1$	18	0.3413	17.07
$>+1$	6	0.1587	7.93

The observed frequency (O) is the actual number of measurements of granite density (standardised). The expected frequency (E) is the number of measurements expected in that z-range for a set of 50 measurements which were normally distributed

$$E = n \times [\text{probability of } z\text{-value occurring in that class interval}]$$

We are now ready to calculate an X^2 value. The null hypothesis is that the sample is from a normal distribution. We can use the usual 5 per cent confidence interval. The X^2 will have $m - k - 1$ degrees of freedom. There are four classes ($m = 4$) and we have estimated $\bar{x}$ and s^2, so $k = 2$. X^2 will have one degree of freedom. This is the minimum possible because if the frequencies in some classes had still been smaller than five, further combination of classes have been impossible (if $m = 3$, degrees of freedom $= 0$). The test procedure would have had to be abandoned and more measurements of granite density collected to produce a reasonable frequency in each class.

From tables of χ^2 values, (for examples, Murdoch and Barnes, 1974, table 8) 95 per cent of the area under the χ^2 curve, with one degree of freedom lies inside a value of 3.84. That is

$$\chi^2_{(0.05;1)} = 3.84$$

The χ^2 distribution with 1 degree of freedom is shown in figure 4.11.

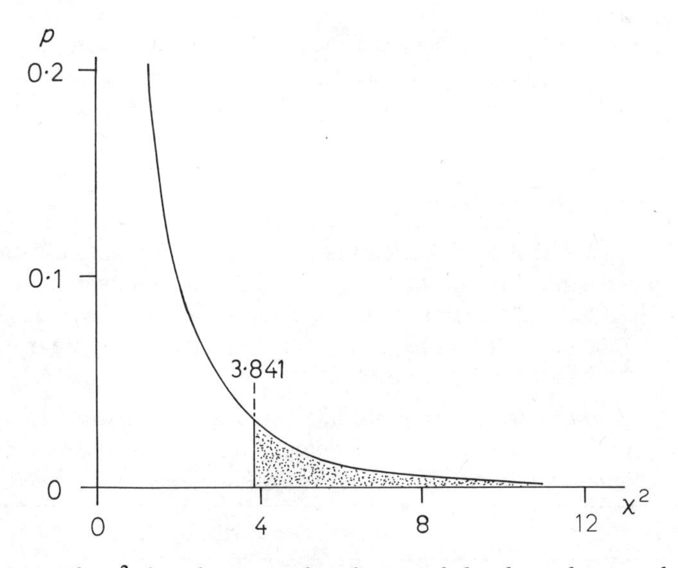

Figure 4.11 The χ^2 distribution with 1 degree of freedom, showing the upper 5 per cent critical region.

If our calculated X^2 value, which follows the χ^2 distribution, is greater than 3.84 we will be able to say that there are less than five in a hundred chances of our set of measurements coming from a normal distribution, and we would reject H_0. Conversely, if we have a normal distribution most samples from it should themselves have something like a normal distribution of frequencies. Therefore the differences between observed and expected values should most often be small, producing a small X^2 value.

We can now calculate our X^2 value, from equation 4.11

$$X^2 = \sum_{i=1}^{4} \frac{(\text{Observed frequency in class } i - \text{Expected frequency in class } i)^2}{\text{Expected frequency in class } i}$$

with $m - k - 1$ d.f.

$$X^2 = \frac{(6 - 7.93)^2}{7.93} + \frac{(20 - 17.07)^2}{17.07} + \frac{(18 - 17.07)^2}{17.07} + \frac{(6 - 7.93)^2}{7.93}$$

$$= 0.470 + 0.503 + 0.051 + 0.470$$

$X^2 = 1.494$ with 1 d.f.

This value is much less than $\chi^2_{(0.05;1)}$ which is 3.84. Consequently we cannot reject H_0 and must conclude that the sample is not significantly different from a normal distribution. There are far more than 5 out of a 100 chances of this sample representing a normal population (actually about 25 out of a 100).

This important χ^2 goodness-of-fit test can be used on any distribution. We might test whether a set of trace element data fits a log-normal distribution or whether a set of cross-bedding directions measured in the field fit a circular-normal distribution.

Test for the Markov property

There is also a statistic, which follows the χ^2 distribution, that can be used to test for the Markov property (section 2.2) in a transition probability matrix. The data in table 4.6 represents a sequence of simplified Coal Measure cyclothems. We erect a null hypothesis H_0, that the successive events (that

Table 4.6 Upward transition probability matrix for sequence of simplified Coal Measure cyclothems. The probability value is in the upper left, and the original tally-matrix frequency in the lower right of each cell.

	Clay		Sand		Seat-earth		Coal		Total	
Clay	0.2	4	0.75	15	0.05	1	0	0	1.0	20
Sand	0.33	12	0.28	10	0.33	12	0.06	2	1.0	36
Seat-earth	0.06	1	0.12	2	0.12	2	0.70	12	1.0	17
Coal	0.25	4	0.56	9	0.13	2	0.06	1	1.0	16
Total	0.24	21	0.40	36	0.19	17	0.17	15		89

is, lithologies) are independent. The alternative H_a, is that they are not independent and hence form a Markov chain.

The statistic λ is calculated

$$\lambda = \prod_{i,j=1}^{m} \left(\frac{p_j}{p_{ij}}\right)^{n_{ij}}$$

where $\prod_{i=1}^{3} x_i$ means the product of

$$x_1, x_2 \text{ and } x_3 = x_1 \times x_2 \times x_3$$

p_{ij} = probability in cell ij (row i, column j) of the probability matrix
p_j = marginal probability in the jth column calculated from column tally divided by total tally
n_{ij} = frequency in cell ij of original tally matrix
m = number of states (rows or columns) in the matrix
and $-2 \log_e \lambda$ follows χ^2 with $(m - 1)^2$ degrees of freedom

$$-2 \log_e \lambda = \sum_{i=1}^{m} \sum_{j=1}^{m} n_{ij} \log_e \frac{p_{ij}}{p_j}$$

which is easier to calculate. For our data in table 4.6, $m = 4$ and

$$-2 \log_e \lambda = 2\left[4 \ln\left(\frac{0.2}{0.24}\right) + 15 \ln\left(\frac{0.75}{0.40}\right) \ldots - 2 \ln\left(\frac{0.13}{0.19}\right) + 1 \ln\left(\frac{0.06}{0.17}\right)\right]$$

$$= 2[- 0.7308 + 9.429 \ldots - 0.7596 - 1.0413]$$

$$= 2[27.0277] = 54.0554$$

$$-2 \log_e \lambda = 54.0554 \text{ with 9 d.f.}$$

$\chi^2_{(0.05;9)} = 16.91$, so we can reject H_0 at a 5 per cent confidence level and conclude that the probability matrix has the Markov property. There are less than 5 in a 100 chances of such a distribution of probabilities arising from successive events that are independent.

4.6 Error theory

In considering the various forms of population distribution, examples have been used where variation was an inherent property of the population. For example, porosity varied throughout the sandstone, and the composition of the volcanic gases varied with place and time. A distribution of values can also be obtained by repeatedly determining some parameter on a single object. The variation in observed value is caused by various random errors which occur during the measuring processes. These errors are called experimental errors and are not to be confused with mistakes. Random errors are inherent in the experimental process, but mistakes involve a departure from the prescribed experimental procedures. Barford (1967, p. 73) says errors arise from 'ambiguities or uncertainties in the process of measurement, or from fluctuations which are too irregular or fast to be observed in detail'. For example, if we can only measure the length of a lamellibranch to the nearest millimetre, anything lying between two values (say 75 or 76 mm) will be recorded as one value of the other. Sometimes we may record 75 mm, but the choice could be random, giving either positive or negative deviations from the real value (that is, errors).

The distribution of errors is usually, and is always assumed to be, a normal distribution. We estimate the parameters of the error distribution to determine the quality of our experimental procedures. A discussion of errors is relevant here, because of the importance of precise and accurate replicate measurements in the fields of geophysics and geochemistry.

Precision and accuracy

These two properties were mentioned in section 1.2. Precision is an estimate of the reproducibility of a method, and is estimated by the standard deviation of the error distribution. This distribution is obtained by analysis of several preparations made from the same specimen—a procedure referred to as replication or analysis of replicates. The better the technique used the more closely the determined values will be bunched about the mean.

If we calculate the mean, $\bar{x}$ and the standard deviation s, of our error distribution, the precision is defined by the coefficient of variation c, where

$$c = \frac{100s}{\bar{x}} \tag{4.13}$$

Using the replicate X-ray diffraction analyses in table 4.7 we find

$$\bar{x} = 25.51$$

$$s = 0.75$$

$$c = 2.92 \text{ per cent} \qquad \text{for } \bar{x} = 25.51$$

The coefficient of variation expresses the standard deviation as a percentage of the mean value and is more informative than just quoting s. The value of c should be quoted for a given mean $\bar{x}$, as above, because s and hence c can vary over the range of possible mean values. For example, figure 4.12 shows a hypothetical working curve for determining the zinc content of a rock by spectrographic analysis. This involves burning the rock in an arc and recording the photographic intensity of the characteristic zinc radiation emitted. Throughout the range of zinc contents we can only determine the measurements on the x-axis with the same degree of distinction, say to the nearest 0.1 of a unit. For low and very high zinc contents a change

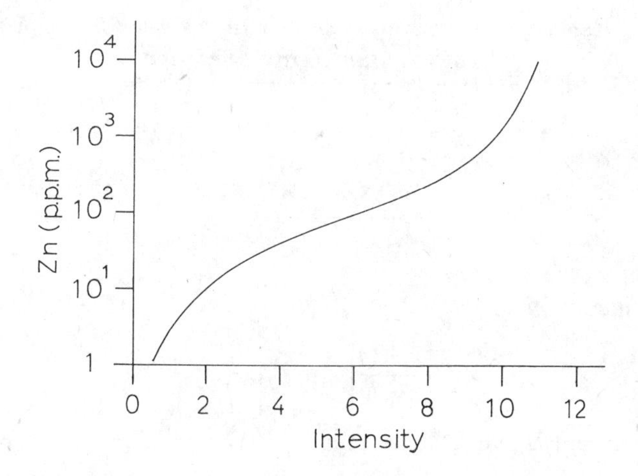

Figure 4.12 A hypothetical working curve for spectrographic analysis

of 0.1 units on the x-axis makes a larger change in the zinc values than in the mid-range of the curve. Consequently the precision obtainable is less and replicate analyses will have a larger spread for the low and high zinc contents.

This type of relationship between precision and the value of $\bar{x}$ is commonly found. Analysts always try and use a technique for which their set of rocks will lie in the mid-range of a working curve, but this assumes some prior knowledge of the composition of the rock. The working curve is produced by analysing a set of standards of known composition or by analysing a set of artificial, prepared standards.

The accuracy of a method is an estimate of how far from the true value is the determined mean value. If a method introduces a systematic error, or bias (for example, an incorrectly zeroed meter) we can produce a precise, but inaccurate result. To estimate the accuracy of a method, a t-test is used to discover if a significant bias is present. To do this we need to know what the true value of our experimentally determined quantity should be for a number of specimens. In some cases we have no way of knowing a true value and must resort to some inter-laboratory experiment (see below) to try and gain an estimate. Sometimes we can make artificial mixtures of known composition, or refer to defined standards, with which to assess the accuracy of our methods.

Suppose we know that the real content of high magnesium calcite in the sample analysed in table 4.7 is 25 per cent. Is our mean value of 25.51 per cent an unbiased estimator of this true value? We erect a hypothesis H_0 ($\mu = 25$) where μ is the true mean value of the population, versus $H_a(\mu \neq 25)$ and calculate a t-value to test H_0.

Table 4.7 Replicate analysis of a carbonate sediment for high magnesium calcite, using X-ray diffraction.

26.3	26.0	25.8
25.1	24.7	24.4
26.8	25.0	25.7
	25.3	

From inequality (4.4)

$$\left|\frac{\bar{x}-\mu}{s/\sqrt{n}}\right| \leqslant t_{(0.05/2;\,n-1)} \tag{4.14}$$

if we can accept H_0 at the 5 per cent level, where

$\bar{x}$ is the mean value of the replicates
μ is the true value for the specimen
s is the standard deviation of the replicates
n is the number of replicates

For the data from table 4.7

$$\left|\frac{\bar{x}-\mu}{s/\sqrt{n}}\right| = \left|\frac{25.51-25}{0.746/\sqrt{10}}\right| = 2.16$$

which is less then $t_{(0.05/2;\,9)} = 2.26$. This means that we must accept H_0 and conclude, at the 5 per cent confidence level, that our determined value does not differ significantly from the true value. There are just more than 5 in a 100 chances of a value of 25.51 arising from a real value of 25 per cent. The method does not have a significant bias. We use a two-tailed t-test in equation (4.14) because bias could be positive or negative.

Propagation of errors

Suppose we are measuring the total thickness of a coastal section of Purbeck rocks which dip along the section at an angle of a few degrees. We will have to add together a number of tape-measure lengths to obtain the total thickness. Each tape-measurement is subject to some random measurement error. These errors will be cumulative in our estimate of the total thickness, and could range from all being positive (an overestimate) to all being negative (an underestimate). Any intermediate positive or negative total error is possible, even to the case where all the errors combine to zero.

As another example, suppose we are measuring the angle of extinction of a felspar grain in a thin section. We read a vernier scale on the microscope stage at the beginning and at the end of rotating the stage. Both of these operations will be subject to a random error which may combine or may cancel each other out, provided that the two errors are independent of each other.

These examples describe two common situations where the maximum possible error is the sum of the individual errors. If we consider the variance of these errors, then we could prove mathematically that the variance of their sum (that is the total error in the required measurement) equals the sum of their variances. If we consider the errors which apply to two measurements x and y, then

$$\sigma^2_{(x+y)} = \sigma^2_x + \sigma^2_y \tag{4.15a}$$

and

$$\sigma^2_{(x-y)} = \sigma^2_x + \sigma^2_y \tag{4.15b}$$

which just say that the error variance of the sum or difference of the two measurements is the sum of their individual variances. Equations (4.15a) and (4.15b) are special cases of the first law of propagation of errors, which in the general case is

$$\sigma^2_{(a \pm b \pm c \ldots \pm n)} = \sigma^2_a + \sigma^2_b + \sigma^2_c + \ldots + \sigma^2_n \tag{4.16}$$

Calculation of our final result in some experimental procedure may involve the multiplication or division of several measured properties. For example, determination of density (δ) requires the measurement of mass M and volume V

$$\delta = M/V$$

The variance of $\delta(\sigma^2_\delta)$ is related to the variance of $M(\sigma^2_M)$ and $V(\sigma^2_V)$ by

$$\left(\frac{\sigma_\delta}{\delta}\right)^2 = \left(\frac{\sigma_M}{M}\right)^2 + \left(\frac{\sigma_V}{V}\right)^2$$

rearranging this in terms of coefficients of variation, using equation 4.13

$$c^2_\delta = c^2_M + c^2_V$$

which leads to a general equation of the second law of propagation of errors. If

$$u = \frac{x \,.\, y \,.\, z \ldots}{p \,.\, q \,.\, r \ldots}$$

then the total error in u, expressed as a coefficient of variation c_u is

$$(c_u)^2 = (c_x)^2 + (c_y)^2 + (c_z)^2 + \ldots + (c_p)^2 + (c_q)^2 + (c_r)^2 + \ldots \tag{4.17}$$

Mandel (1964), Barford (1967) and Clifford (1973) all describe and calculate many types of experimental errors.

Inter-laboratory experiments

To assess the accuracy of a method of measurement for some property, we need to analyse some object, called a standard, with a known value for that property. Imagine we are using some method of measuring mass; in theory we can assess its accuracy by employing the method on the international prototype of the kilogram. It is rather unlikely that we would be allowed to do this directly, but there are definite standards available for all SI base units (see Royal Society, 1971, p. 21–2). The lack of such standards faces measurers in many parts of science, and has been overcome by prescribing specimens of some unknown for analysis by many different people in an inter-laboratory experiment. A mean result is taken as the accepted value for the measurement—a sort of standardisation by consensus. The remainder of the material then becomes available as a standard.

When performing a chemical analysis on a rock there are no primary standards available by which to test our methods. Artificial standards could be prepared, but they do not possess the same characteristics as a real rock. For example, alumina (Al_2O_3) might be used in an artificial standard, whereas aluminium would be present in an igneous rock in a complex silicate mineral. Though the two compounds may contain the same amount of aluminium, their chemical behaviour would be quite different. Several inter-laboratory experiments in geochemical analysis have been mounted with varying degrees of lack of success. The first of these projects was co-ordinated by the *United States Geological Survey* and is reported in Fairbairn (1951), and Stevens and Niles (1960). At this time most analysts used adaptations of the same wet chemical techniques.

Large specimens of a granite (a six-feet (1.8 m) long by six-inch (150 mm) square block) and a diabase (260 pounds (120 kg) of large lumps) were ground to a fine powder, carefully homogenised, and split and bottled in 70 gram amounts. The powders, referred to as G-1 and W-1 were sent out to some forty analysts all over the world. Their whole rock analyses (quoting percentages of SiO_2, TiO_2, Al_2O_3, Fe_2O_3, FeO, MnO, MgO, CaO, Na_2O, K_2O, P_2O_5, CO_2, $H_2O(+)$ and $H_2O(-)$) were treated statistically to obtain mean values for all these constituents. It was hoped that laboratories all over the world could then use the remaining bottles of G-1 and W-1 as reference standards. Unfortunately the different analytical techniques and the different laboratories produced quite different values for these major chemical constituents and different elements were reported with very different levels of precision. For example, the coefficients of variation reported for G-1 ranged from 32 per cent for Fe_2O_3 to 0·26 per cent for SiO_2 (Stevens and Niles, 1960, table 3).

Naturally geochemists were worried by the outcome of this project. Firstly, rock chemical analysis was less precise than they thought, and secondly the comparison of sets of data produced by different workers was realised to be unreliable because of the apparent bias produced by different techniques. Over the years many analyses of G-1 and W-1 were reported in the literature and summarised (Fleischer and Stevens, 1962; Fleischer, 1965). Recommended values for the major constituents and many trace elements were defined, but many problems were raised relating to the initial design of the inter-laboratory experiment. For example, Vistelius (1970) suggests that the original powders despatched to analysts may have been bottled from inhomogeneous powder. Since the G-1/W-1 project, many more standard rocks have been prepared, despatched and analysed. The results of these exercises are regularly published (for example, Flanagan, 1973).

Chayes (1970, p. 207) summarises the effects of geology's inter-laboratory experiments.

> One of the most disappointing features of the G-1/W-1 program was that it did not lead to a serious attempt to identify the various sources of significant error in classical silicate analysis and evaluate the contribution each makes to the error variance. It is difficult to appreciate now the combination of apathy and antipathy with which most silicate chemists and petrologists alike then viewed the whole subject of statistical analysis....
>
> Considering the situation as a whole, the G-1/W-1 program was immensely informative. If its first impact had been felt in the late 50s instead of nearly 10 years earlier, the result might well have been a properly designed inter-laboratory study of the error of silicate analysis.

It is to be hoped that similar exercises will be more fruitful in the future.

4.7 Applications

The t, F and χ^2 tests are often used as part of a more complex statistical procedure and they will crop up repeatedly in the following chapters.

Koch and Link (1963) use t-values to erect confidence intervals on heavy metal contents of an ore vein on either side of a fault. Doornkamp and King (1971, pp. 38–41) use a t-test to show a significant difference between basin areas in two different drainage basins. They (*ibid.*, pp. 310–16) also use a t-test to show that glacial moraines have significantly different heights in different parts of a glacial region. Hayami (1973) uses t-tests as part of a statistical analysis of samples of a Pliocene-recent scallop species from Japan. He is able to establish the presence of several phenotypes. Reyment (1966) employs t-tests and confidence intervals to analyse the form of holes drilled by a boring gastropod in the shells of a pelecypod found on the Niger delta.

Krinsley (1960) compares the mineralogy (aragonite content) and chemical composition (magnesium and strontium content) of three gastropod genera. Prior to the use of a t-test, he uses an F-test to show that the variance of his measurements on each genus is not significantly different.

Bozdar and Kitchenham (1972) studied the distribution and productivity of lead mines on the Alston Block in the North Pennines, England. They use χ^2 goodness-of-fit tests to show:

1. that the geographical distribution of mines follows one type of binomial distribution.
2. that the productions of the lead mines are log-normally distributed.

From these conclusions predictions are made for the discovery and working of mines on the Askrigg Block, which lies to the south of the Alston Block. Harbaugh and Bonham-Carter (1970) and Read (1969) use and discuss the χ^2 statistic in testing sedimentary sequences for the Markov property.

5 CORRELATION AND REGRESSION

Most students have heard of the statistical methods of correlation and regression, even if they do not know how to perform the calculations. Simple linear correlation assesses the degree of relationship between two variables measured on a number of individuals. Regression allows us to quantify the relationship between this pair of variables (if a significant relationship does exist), so that other values can be predicted.

It must be stressed that these statistical tests can be used on any pair of variables. We must make the decision to correlate them, because we believe that they are related to each other, or that some control is governing their behaviour. It is easy to calculate highly significant but 'nonsense' correlations.

There is a tendency to bundle raw data straight into computer programs to calculate correlation coefficients for a large number of variables on large samples. Conclusions are then based on the resulting correlation coefficients, without ever looking at a graph of the variable pairs. The graph should always be plotted and used when interpreting correlation coefficients and regression lines.

With this cautionary note, it must be stated that correlation and regression are powerful tools. When used properly they form the basis for many further statistical procedures. (Rank correlation is described in chapter 7.)

5.1 Correlation

Pearson's product-moment coefficient of linear correlation

Often we want to ask questions of the sort: 'Is particle size related to particle shape in a beach sand?'; 'Do potash and alumina vary together in this lava suite?'; 'Are length and breadth related in this mollusc species?'; or 'Does clay content vary linearly through this stratigraphic sequence?' To answer these questions we can use Pearson's product-moment coefficient of linear correlation, normally referred to as r, the correlation coefficient, to assess the linear relationship between a sample of x and y values we have measured.

The sample correlation coefficient r (or $\hat{\rho}$), which is an estimate of the population correlation coefficient ρ, is given by

$$r = \frac{\text{covariance}\,(x, y)}{\sqrt{[\text{variance}\,(x) \times \text{variance}\,(y)]}}$$

where n values of $x(x_1, x_2 \ldots x_n)$ and of $y(y_1, y_2 \ldots y_n)$ were obtained. That is

$$r = \frac{\dfrac{\sum (x - \bar{x})(y - \bar{y})}{n - 1}}{\left[\dfrac{\sum (x - \bar{x})^2}{n - 1} \times \dfrac{\sum (y - \bar{y})^2}{n - 1}\right]^{\frac{1}{2}}} \tag{5.1}$$

where $\bar{x}$ and $\bar{y}$ are the mean values of x and y. This formula is the ratio of how much x and y vary together about their means to the total variation of x and y. However the correlation coefficient can be expressed in a form which is much simpler to compute

$$r = \frac{\text{CSCP}}{\sqrt{(\text{CSSX}\,.\,\text{CSSY})}} \tag{5.2}$$

where CSCP (corrected sum of cross products) $= \sum xy - \sum x\,.\sum y/n$
and CSSX (corrected sum of squares of x) $= \sum x^2 - \sum x\,.\sum x/n$
and CSSY (corrected sum of squares of y) $= \sum y^2 - \sum y\,.\sum y/n$

where

$$\sum_{i=1}^{n} x_i = \sum x \text{ etc.}$$

It is easy to check that equations (5.1) and (5.2) are equivalent. Equation (5.2) requires the calculation of five different sums; the sum of xs, sum of ys, sum of x^2s, sum of y^2s and the sum of the xy values. The computations can be done on a calculating machine, though some would consider this to be unnecessarily slow in this day of fast computers. However all students should at least start by calculating a few values of r by hand.

Values of r, which is a dimensionless measure, can vary between $+1$ and -1. $r = +1$ expresses perfect sympathy between x and y. That is x and y have a perfect linear relationship and covary together. $r = -1$ expresses perfect antipathy between x and y and $r = 0$ means that there is no relationship between x and y. Figure 5.1 shows some graphs for various values of r.

Table 5.1 gives a set of measurements of basin area and total stream lengths in a drainage system. Empirically we may suspect that there is a relationship between these two variables of a drainage system. Initially this may be an intuitive feeling resulting from our geomorphological observa-

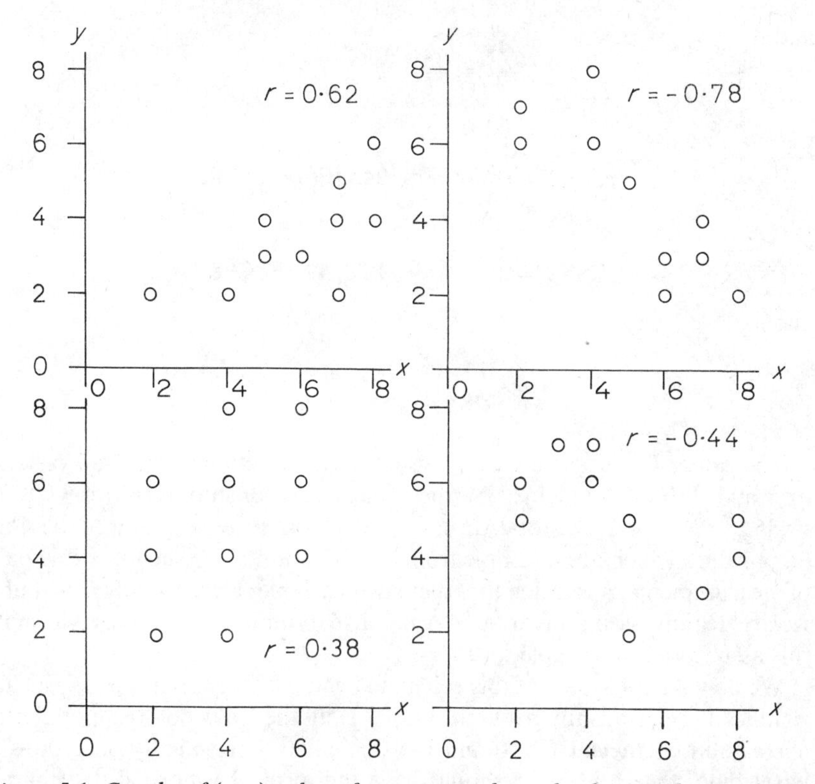

Figure 5.1 Graphs of (x, y) values for various values of r the sample correlation coefficient.

tions, but we can test this relationship by calculating the correlation coefficient. From table 5.1

$$n = 10$$

$$\sum x = 216.0 \qquad \sum x^2 = 4993.66$$

$$\sum y = 113.6 \qquad \sum y^2 = 1377.48$$

Table 5.1 Values of basin area and total stream length for a drainage system. Measurements in kilometres and square kilometres.

Stream length	Basin area	Stream length	Basin area
12.0	6.3	26.3	11.7
16.4	7.3	23.0	12.5
15.8	9.5	25.4	13.6
21.0	10.5	28.6	14.5
17.5	11.4	30.0	16.3

and

$$\sum xy = 2609.77$$

$$\bar{x} = 21.6 \qquad \bar{y} = 11.36$$

$$\text{CSCP} = 2609.77 - 2453.76 = 156.01$$

$$\text{CSSX} = 4993.66 - 4665.6 = 328.06$$

$$\text{CSSY} = 1377.48 - 1290.49 = 86.98$$

and

$$r = \frac{156.01}{(328.06 \times 86.98)^{\frac{1}{2}}} = 0.9235$$

The value of r^2 is also useful, for it tells us the fraction of the total variance of x and y that is explained by their linear relationship. Here $r^2 = 0.8528$, or 85.28 per cent of the total variance. Figure 5.2 shows a graph of these ten (x, y) values which are a sample from a population of all such possible pairs of measurements. Assuming that these two variables are normally distributed we are dealing with a bivariate normal distribution. It is necessary to make this assumption to be able to test r.

We now want to know if this calculated value of r represents a statistically significant relationship between x and y. If the value of the population correlation coefficient $\rho = 0$ and if we calculate r repeatedly on groups of basin data, the values of r should have the form of a normal distribution

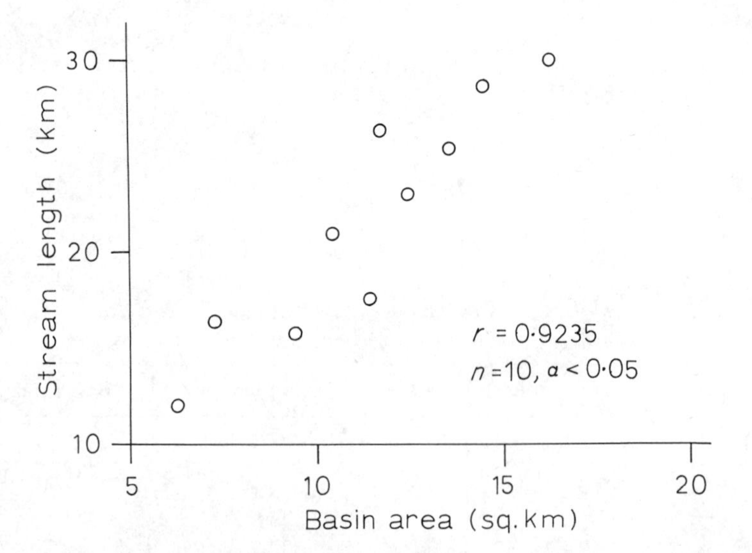

Figure 5.2 Graph of basin area against total stream length for a drainage basin.

with a mean of zero. Because this is so we can test our value of r against a hypothesis $H_0(\rho = 0)$ using a t-test

$$t = r\sqrt{\left(\frac{n-2}{1-r^2}\right)} \qquad \text{with } (n-2) \text{ d.f.} \tag{5.3}$$

We use a one-tailed t-test because out alternative hypothesis is $H_a(\rho > 0)$, since we are interested in establishing a relationship between basin area and stream length. (If r were negative the hypothesis would be $H_0(\rho = 0)$ versus $H_a(\rho < 0)$ and again we would use a one-tailed t-test.)

For our data

$$t = 6{\cdot}81 \qquad \text{with 8 d.f.}$$

Using a one-tailed Student's t-distribution at the 0.05 level, we can reject H_0 if $t > t_{(0.05;8)} = 1.86$. This means that our t-value lies in the top 5 per cent of the t-distribution. So there are far less than five in a hundred chances of obtaining such an extreme sample correlation coefficient r, from a population coefficient of ρ with a true value zero.

If ρ is not equal to zero, then the sampling distribution of r is not normal, but is skewed (see Fryer, 1966, p. 228, for experimental proof). To be able to apply tests to such values of r we use the Z-transformation that was proposed by Fisher

$$Z = \log_e \sqrt{[(1+r)/(1-r)]} = 1.1513 \log_{10}\left(\frac{1+r}{1-\text{r}}\right) \tag{5.4}$$

Z has an approximately normal distribution with the following properties

$$\mu_Z = 1.1513 \log_{10}\frac{(1+\rho)}{(1-\rho)}$$

$$\sigma_Z = \frac{1}{\sqrt{(n-3)}} \tag{5.5}$$

Note that σ_Z depends only on sample size. Having produced a statistic with a normal distribution we can now use our standard tests on it. We can test the value of r from the drainage basin data to see if it is drawn from a population with ρ as small as 0.5. So if $\rho = 0.5$

$$\mu_Z = 1.1513 \log_{10}\left(\frac{1+0.5}{1-0.5}\right) = 0.5493$$

$$\sigma_Z = \frac{1}{\sqrt{(n-3)}} = 1/\sqrt{7} = 0.3780$$

$$Z = 1.1513 \log_{10}\left(\frac{1+0.9235}{1-0.0235}\right) = 1.612$$

From this a standardized z is calculated

$$z = (Z - \mu_Z)/\sigma_Z = 2.81$$

whose value we can look up in the table of areas under the normal curve (table 3.6). Our value of z tells us that 99.75 per cent of the normal curve lies inside the value $z = \pm 2.81$. Consequently our value of $r = 0.923$ could occur 0.25 per cent of the time from a population whose true $\rho = 0.5$. Normally we use our 5 per cent significance level (or $\alpha = 0.05$). Therefore we can at this level reject the hypothesis that our value of r represents a ρ which is as small as 0.5. This means our value of r is from a population whose ρ is significantly greater than 0.5.

Alternatively we can compare two correlation coefficients to see if they are significantly different by calculating

$$z = \frac{Z_1 - Z_2 - (\mu_{Z_1} - \mu_{Z_2})}{\sqrt{(\sigma_{Z_1}^2 + \sigma_{Z_2}^2)}} \tag{5.6}$$

In practice most people just refer to a table of values of r to assess its significance. These were first published by Fisher and Yates (1948) and are found in most statistical tables (see Murdoch and Barnes, 1974, table 10). For a given number of degrees of freedom $(n - 2)$ a series of significant values of r are tabulated ($\alpha = 0.05, 0.01$, etc.). These tables give us a minimum value of r for rejecting the hypothesis that $\rho = 0$, at that α level of significance.

Graphing variables

In the introduction, the importance of actually plotting graphs (or scatter diagrams) of the samples of x and y values was stressed. This is particularly important when interpreting the correlation coefficient.

If the sample data being analysed contains one extreme value, then this single point can produce a large value of r. Figure 5.3 shows this most clearly for some artificial data. When collecting the data for statistical analysis we must try to avoid such samples.

The following example underlines the point that we must decide on the questions that our statistical tests are meant to answer. A statistic will give some sort of answer to any question.

Figure 5.4 shows the relationship between calcium and magnesium content for recent carbonate sediments from Belize. The total sample shows a significant positive correlation ($r = 0.5512$, $\alpha < 0.05$). This can be explained from our knowledge of the presence of magnesium in calcites. In looking at the graph the different environments seem to plot as separate groups. Indeed if we split the sample into two sub-samples; one—the reef and back-reef samples—shows a significant negative correlation between calcium and magnesium; the other—the lagoon and near-shore samples—shows a positive correlation. Again we can explain these differences from geological

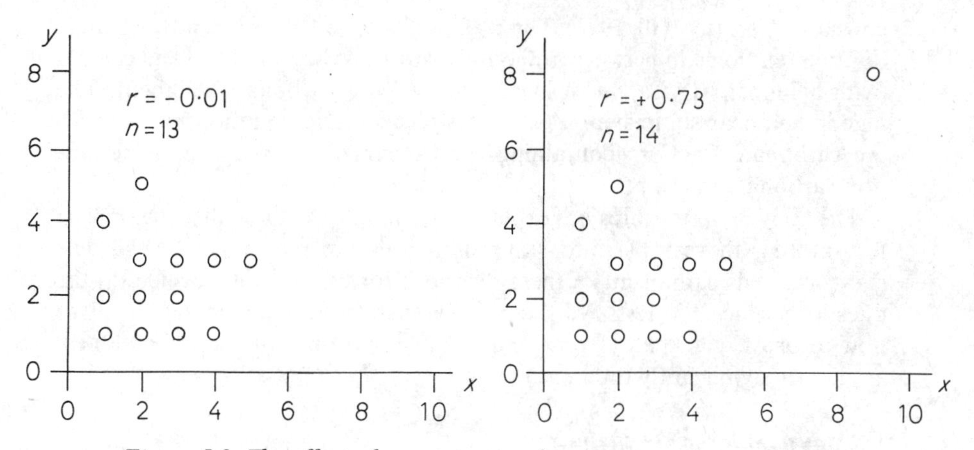

Figure 5.3 The effect of an extreme value on the r-value computed.

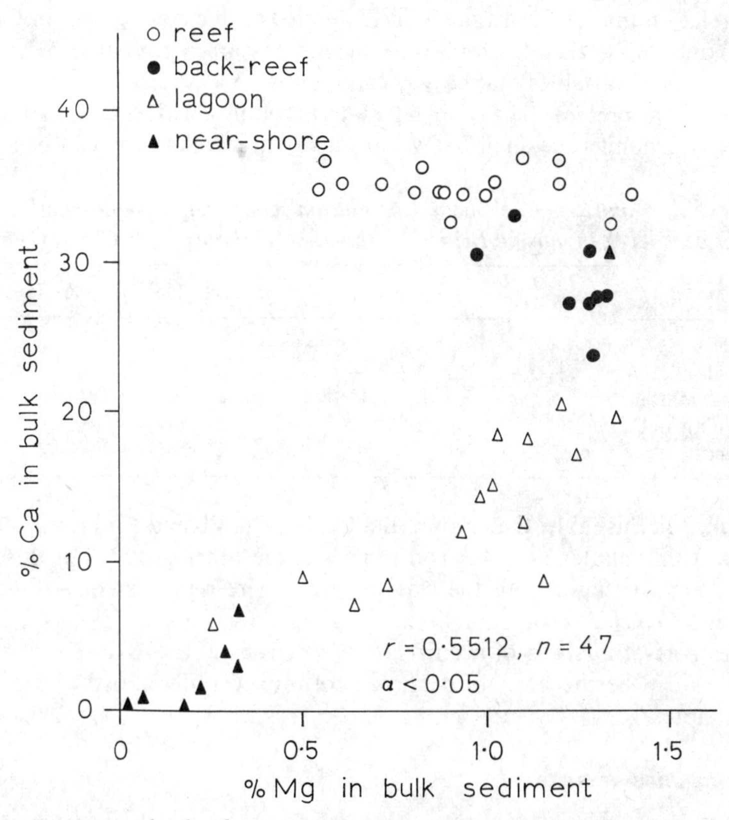

Figure 5.4 A graph of calcium versus magnesium content for recent sediments from south of Belize City, Belize (from Till, 1971, p. 527).

considerations (see Till, 1971). The reef and back-reef samples show a simple dilution relationship between aragonite (with a low Mg/Ca ratio) and calcites (with a higher Mg/Ca ratio) in an almost pure carbonate sediment. The lagoon and near-shore samples show a similar dilution relationship between the carbonate fraction (containing the calcium and magnesium) and the non-carbonate fraction.

The above relationships are explained quite fully to show that the criteria for splitting the sample into sub-populations are purely geological. The statistical tests cannot make these decisions for us. We must decide on the questions which the tests will help us to answer. (In chapter 6 we will see how to prove that the different sedimentary environments described above are in fact significantly different.)

Beginning multivariate analysis

Having measured a large number of variables on a suite of samples, it is useful to calculate correlation coefficients for all variable pairs and to erect a correlation matrix from this data. Examples of this can be seen in Curtis (1969) and Till (1971). If significant correlations are represented by +s or −s, groups of variables that covary can readily be identified.

Table 5.2 represents part of such a matrix used by Curtis (*ibid.*) in studying British Carboniferous shales. Two groups of trace elements can easily be

Table 5.2 Major/trace element correlation coefficients significant at the 5 per cent level, in marine British Carboniferous shales, from Curtis (1969).

	Ca	B	Cr	V	Sr	Ba	Co	Ni	Cu	Pb	Mn
Al_2O_3	+	+	+	+	+						−
Total alkalies	+	+	+	+	+						−
Organic matter							+		+	+	
Total diagenetic Fe								+	+		+
Sulphur						+	+	+	+		

identified in this table. One group (Ba, Co, Ni, Cu, Pb and Mn) is associated with organic matter and the iron minerals; the other group (Ca, B, Cr, V and Sr) is associated with the clay minerals (here represented by the constituents Al_2O_3 and total alkalies).

This sort of analysis of a correlation matrix is not only useful in itself, but also can be the basis of a number of multivariate statistical methods (see chapter 8).

Closed number systems

A whole-rock chemical analysis commonly consists of the ten oxides SiO_2, TiO_2, Al_2O_3, Fe_2O_3, FeO, MnO, MgO, CaO, K_2O and Na_2O. In addition

$H_2O(+)$, P_2O_5 and perhaps CO_2 are determined; for example see Bayly (1968). These oxides are quoted as percentages and a respectable analysis should sum to between 99.5 and 100.5 per cent.

Chayes (1960) first pointed out that this constancy of sum leads to constraints on the correlations between oxides. Such data is referred to as a closed number system.

If a silica value is determined then the next oxide is not free to take any value, but is restrained to be at most (100 – silica). The next oxide is again restrained to be less than (100 – silica – second oxide). The tables of significant values of r were computed for open-number systems in which both variables are free to take any value.

As one would expect in a closed-number system, there is a tendency for negative correlations to be enhanced. This means that if significance is to be attached to a negative value of r, then it must be a great deal more significant than the tables or a t-test would suggest. Similarly, positive values of r could be smaller than the tests would suggest. Chayes and other workers have attempted to calculate these significant values of r for closed-number systems. The mathematics involved is complex and this leads most geologists to ignore their work. Attitudes are improving and Chayes (1971) has devoted a book to the problems of closed-number systems and ratio correlation, and Koch and Link (1971, chapter 11) discuss various approaches to the problems. However, no significance tables for r values in a closed-number system have as yet been calculated.

5.2 Classical regression

The value of r may show that a statistically significant linear relationship exists between two variables. In some cases we may want to quantify such a relationship. For example, table 5.3 gives a set of bulk density measurements made down a core of clay. Is there a linear change in bulk density down the core and how can we quantify the rate of change (or compaction)? The data listed in table 5.3 are plotted in figure 5.5.

There are many ways in which we could fit a line-of-best-fit to any set of data points, but most commonly we apply a least-squares fit, as shown in figure 5.6.

The line constructed in figure 5.5 is the classical regression of variable y on variable x, and is drawn such that the sum of the squares of the vertical deviations about the line is a minimum. In this way a maximum part of the variability of x is explained by the regression line.

The equation of the population line from which our clay specimens are a sample can be expressed as

$$y = \delta + \gamma x + \varepsilon$$

whose δ is the intercept, γ the slope and ε is a random-error term which

Table 5.3 Bulk density (g/cm^3) variation with depth (cm) in a core of clay from the Gulf of Maine. Data provided by A. Parker.

Depth	Bulk density	Depth	Bulk density
1.25	1.36	71	1.41
11	1.34	81	1.415
21	1.38	91	1.43
31	1.40	101	1.43
41	1.43	111	1.44
51	1.40	121	1.42
61	1.40		

produces the variation of the y values about the line. ε is distributed normally with a mean of zero and a variance of σ^2. Consequently we expect to obtain a normally distributed series of values of y, on repreatedly measuring it for a given x-value.

We are saying that the value of y depends on the value of x and we perform a regression of y on x. This means that we recognise an independent variable x and a dependent variable y. The former should be known without error and the latter has a random-error term (ε) applicable to it. In practice we hope that the error on x is orders of magnitude smaller than that on y. In our example, the determination of bulk density of a clay is subject to experimental errors which cause a random fluctuation about the true value, but we can measure the distance down a core with far greater precision.

To use a classical regression line we must believe that one variable depends on the other and ideally should have one variable subject to negligible error.

We will try to estimate the population regression line by computing a sample regression line of our clay data

$$y = a + bx \tag{5.7}$$

with a variance of ε called s^2. Where a is the sample estimate of δ the intercept, b is the sample estimate of γ the slope, and y is the estimated average y (predicted y) for the given x.

Figure 5.6A shows the total sum of squares of the dependent variable y, about its mean $\bar{y}$, which is $\sum_{i=1}^{n} (y_i - \bar{y})^2$.

Empirically we can see that the better the 'fit' of the regression line to the data, the nearer all the points will be to the fitted line, the greater will be the sum of squares associated with the line (figure 5.6B) and the smaller will be the sum of squares of the deviations about the line (figure 5.6C).

The derivation of the least-squares regression line is given in Spiegel (1961, appendix VIII) and will not be produced here. The sample linear

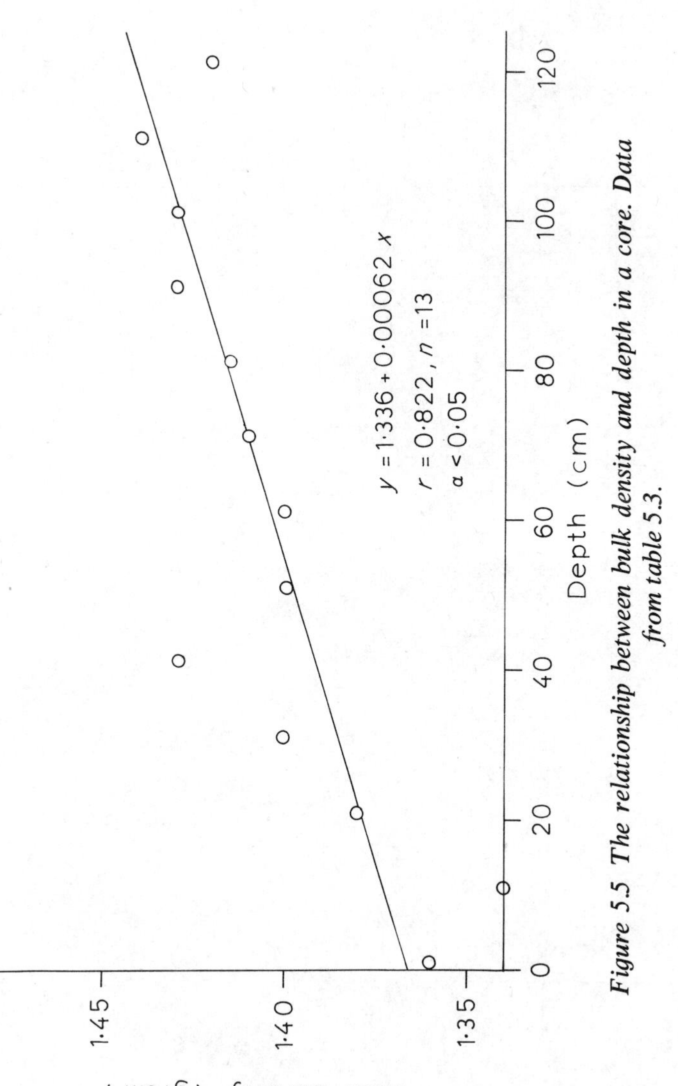

Figure 5.5 The relationship between bulk density and depth in a core. Data from table 5.3.

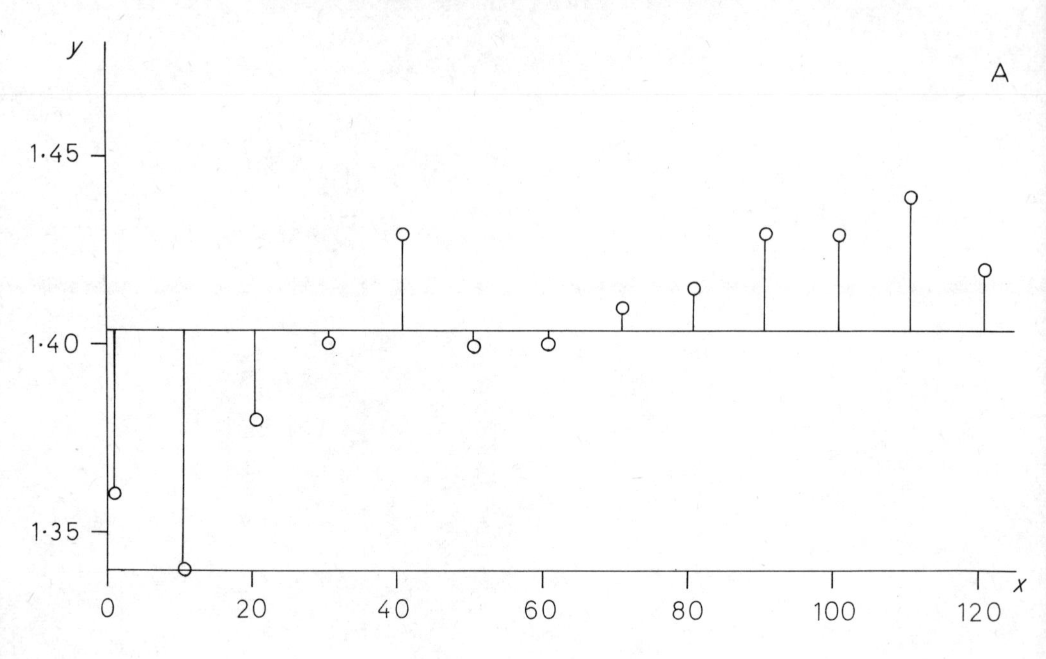
y
A
1·45
1·40
1·35
0
20
40
60
80
100
120
x

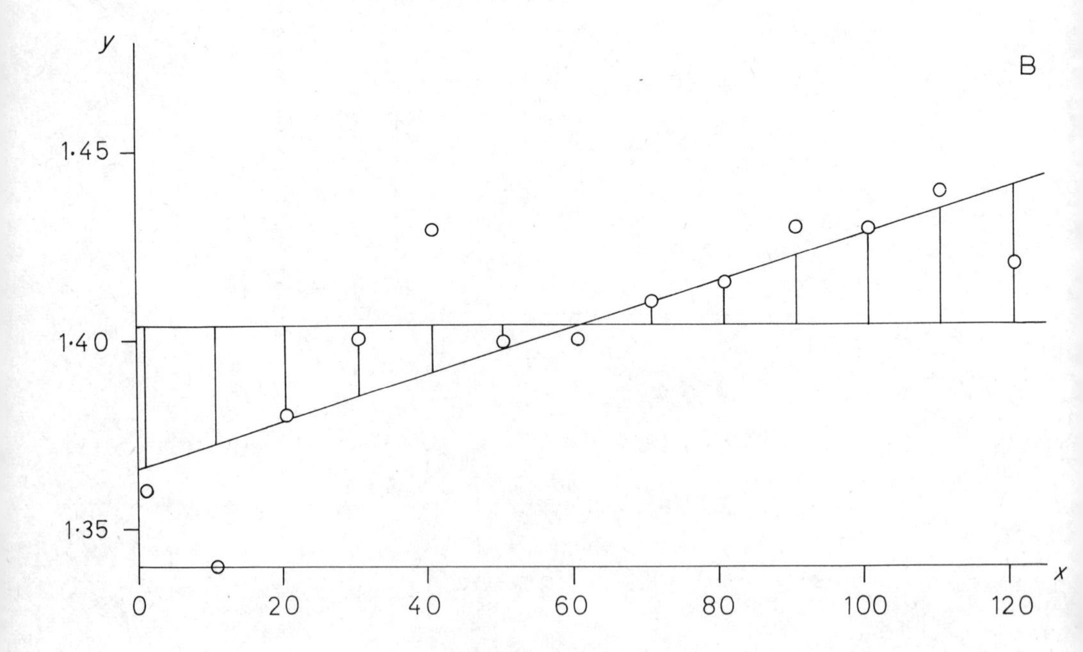
y
B
1·45
1·40
1·35
0
20
40
60
80
100
120
x

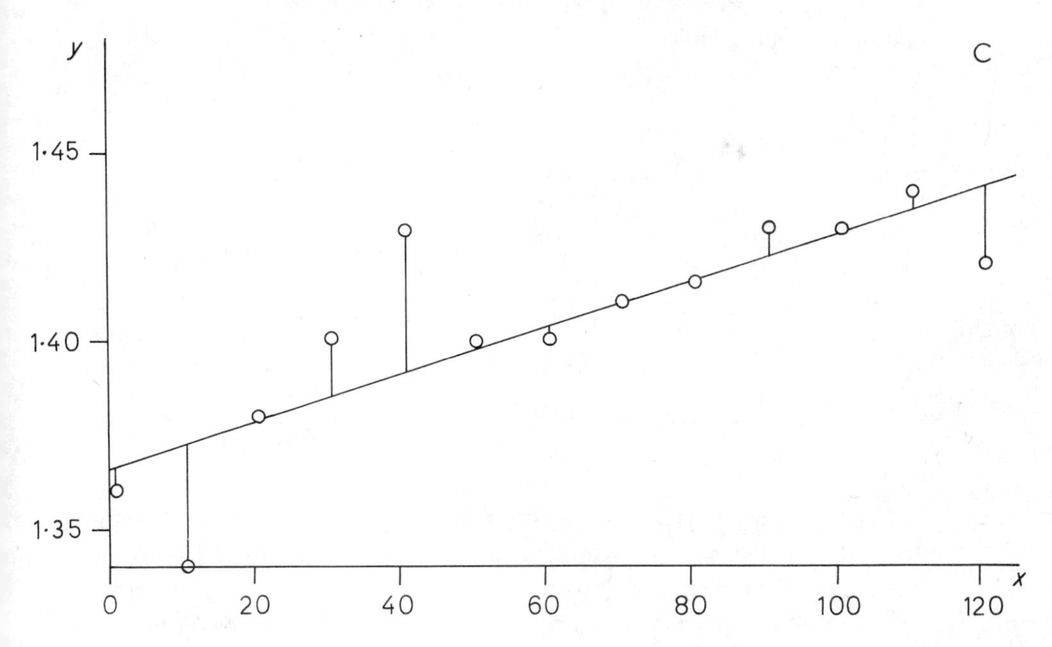

Figure 5.6 Graphical explanation of the classical regression of y on x: A—Total sum of squares of y about its mean $\bar{y}$; B—Sum of squares of y associated with the regression; C—Sum of squares of the deviations of y from the regression.

regression line can be expressed as

$$y = a + bx$$

with a variance of s^2 for ε. Using the notation of section 5.1

the slope $$b = \frac{\sum xy - \sum x \sum y/n}{\sum x^2 - (\sum x)^2/n} = \frac{\text{CSCP}}{\text{CSSX}}$$

the intercept $$a = \sum y/n - b\sum x/n = \bar{y} - b\bar{x}$$

the variance $$s^2 = \frac{1}{n-2}\left[\text{CSSY} - \frac{(\text{CSCP})^2}{\text{CSSX}}\right]$$

It is very easy to show that this regression line goes through the point $(\bar{x}, \bar{y})$ which is the centroid or centre of gravity of the sample of points.

For our data in table 5.3

$$n = 13,$$

$$\sum x = 793.25 \qquad \sum x^2 = 66573.6$$

$$\sum y = 18.255 \qquad \sum y^2 = 25.6446$$

$$\sum xy = 1125.2$$

$$\bar{x} = 61.02 \qquad \bar{y} = 1.404$$

$$\text{CSCP} = 11.2887$$

$$\text{CSSX} = 18170.1$$

$$\text{CSSY} = 0.01038$$

$$r = 0.822 \qquad \text{with 11 d.f.}$$

This value of r is significant at the 5 per cent level ($r_{(0.05,11)} = 0.5529$). This tells us that we are justified in proceeding to calculate a regression line to quantify the relationship.

The value of r is quite high ($-1 \leqslant r \leqslant +1$), but it is a sobering fact to find that $100r^2$ is only 67.6 per cent. This is the amount of the variance of y, the bulk density, that is explained by a linear $x - y$ relationship. The value of r can be employed in this way to assess the usefulness of the classical regression line.

Calculating the regression line we find

$$b = 11.2887/18170.1 = 6.213 \times 10^{-4}$$

$$a = 1.404 - (6.2 \times 10^{-4}) \times 61.02 = 1.366$$

$$s^2 = \frac{1}{12}\left[0.01038 - \frac{(11.2887)^2}{18170.1}\right] = 3.06 \times 10^{-4}$$

which gives

$$y = 1.366 + 0.00062x \qquad (5.8)$$

This is our sample regression line which is an estimate of the population regression line. If we repeatedly collected samples of 13 clay specimens from the cores in the Gulf of Maine and calculated the regression line for each set of samples we would end up with a bundle of lines of varying slope and intercept (b and a respectively). The values of a and b would be normally distributed about δ and γ the population parameters. We need to obtain confidence intervals for the true δ and γ so that we can perform tests on our sample line. For example, we may want to compare the regression line of equation (5.8) with the form of the sample line for another clay, to discover if the clays have different compaction behaviours.

The confidence interval on a, with confidence level $1 - \alpha$ (for example, at the 95 per cent level, $\alpha = 0.05$) is

$$a - t_{(\alpha/2;n-2)} \sqrt{\left(\frac{s^2 \sum x^2}{n\,\mathrm{CSSX}}\right)} < \delta < a + t_{(\alpha/2;n-2)} \sqrt{\left(\frac{s^2 \sum x^2}{n\,\mathrm{CSSX}}\right)}$$

The confidence interval on b is similarly

$$b - t_{(\alpha/2;n-2)} \sqrt{\left(\frac{s^2}{\mathrm{CSSX}}\right)} < \gamma < b - t_{(\alpha/2;n-2)} \sqrt{\left(\frac{s^2}{\mathrm{CSSX}}\right)}$$

Now $t_{(0.05/2;11)} = 2.201$, so using all the numbers we calculated for equation (5.8)

the 95 per cent confidence interval on the intercept a = 1.366 ± 0.02
the 95 per cent confidence interval on the slope b = 0.00062 ± 0.00028

Suppose measurements of bulk density on another core in another area gives the regression

$$y = 1.4 + 0.001\,x \tag{5.9}$$

From above, the value of the slope is outside the 95 per cent confidence limit for our first population line. Therefore we can consider the slopes of the two lines (5.8) and (5.9) to be significantly different, because there are less than 5 in a 100 chances of them both coming from the same population of slopes.

This procedure leaves us with a clearly formulated geotechnical question—which may have been suspected when we sampled originally—namely why is there a different compaction rate in the two clay cores? Statistics offer no help here.

If we use the regression line to predict values of bulk density at a given distance down the core, we need to know a confidence limit for the true population **y**-value, at x_i, from our predicted y.

The confidence interval on y at the $1 - \alpha$ level is

$$y - t_{(\alpha/2;n-2)}\, s \sqrt{\left(\frac{1}{n} + \frac{(x_i - \bar{x})^2}{\mathrm{CSSX}}\right)} < \mathbf{y} < y - t_{(\alpha/2;n-2)}\, s \sqrt{\left(\frac{1}{n} + \frac{(x_i - \bar{x})^2}{\mathrm{CSSX}}\right)}$$

If we calculate this confidence interval for a number of x values we can draw a confidence band on the regression line, as shown in figure 5.7. This confidence band will always be wider at the ends of the fitted line than at the centre, because there are more points near the mean values. Similarly the error in estimating the slope increases as distance from the mean values increases.

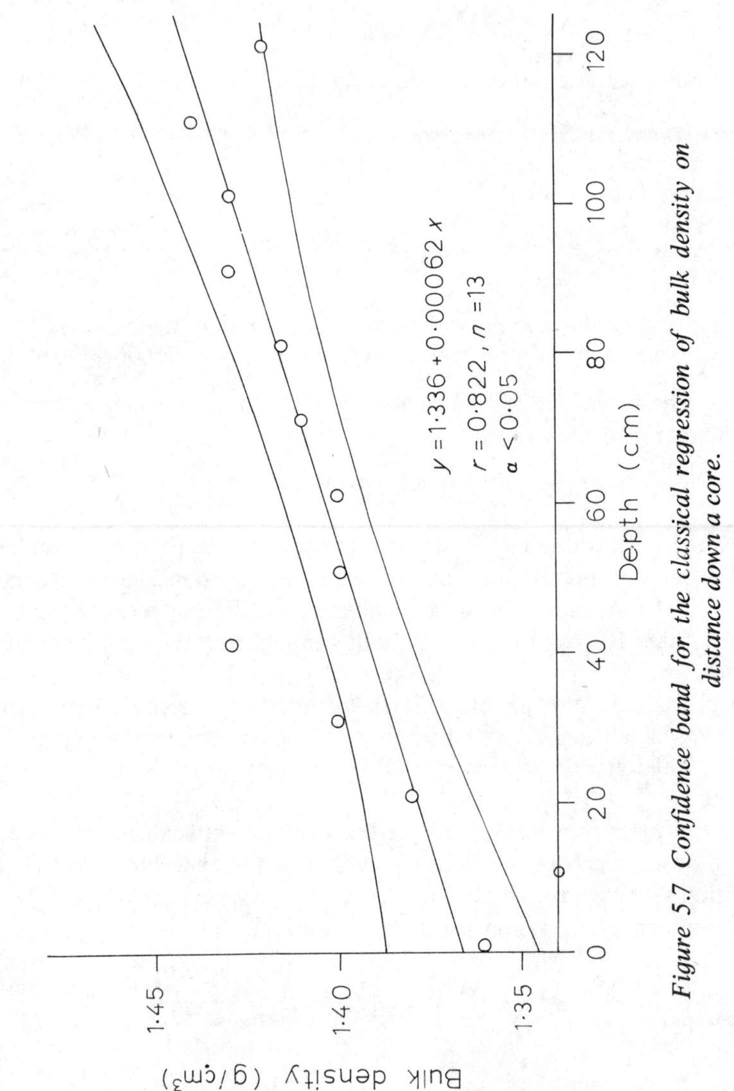

Figure 5.7 Confidence band for the classical regression of bulk density on distance down a core.

5.3 The reduced major axis line

In our regression example above there is a clearly defined dependent variable, for bulk density of the clay is a function of, and dependent on, distance down a core. In some cases the dependence may not be so obvious.

Table 5.4 gives some measurements of width and length of tests of *Nucleolites scutatus.* The correlation coefficient for this data is $+0.9770$, indicating a significant positive correlation at the 5 per cent level. The regression line would be useful for defining this species, but which is the dependent variable? We have no evidence or belief that either variable depends directly on the other, both probably depend on some other vital organic control. We could calculate both classical regressions of y on x, and x on y. These will show a wide divergence. It is worth the exercise to calculate them. In fact the larger the value of r, the correlation coefficient, the nearer together are the two regressions. In the ultimate case where $r = 1$, the regressions of y on x, or x on y become one and the same line.

If we are exploring the relationship between potash and alumina in a lava suite we would not believe that one variable is dependent on the other, rather that both are associated with a particular mineral phase. Both elements will have been determined by some chemical or X-ray technique and will be subject to errors of the same order of magnitude.

In both the above examples we cannot employ classical regression to quantify a linear relationship because the requirements of (1) dependency and (2) knowing one variable without error, cannot be met. To overcome this situation Kermack and Haldane (1950) developed the reduced major

Table 5.4 Measurements (mm) on tests of the echinoid Nucleolites scutatus *collected from the Jurassic at Shellingford and Sheepstead, England. Data provided by R. Sawyer.*

Length (x)	Width (y)	Length (x)	Width (y)	Length (x)	Width (y)
11	11	20	21	24	23
15	11	20.5	20.5	25	24
16	15	21	19	26	24
16	16.5	21	20	26	24
16.5	16	21	20	26	25
18	15	22	20	26	26
18.5	16	22	20.5	28	27
18.5	17	22	21	28.5	26
18.5	17.5	22	21.5	29.5	28.5
18.5	17.5	22.5	21.5	30	28
19.5	18.5	23	21	30.5	28
19.5	19.0	23	22	31	29
20	19	23.5	21		
20	19	23.5	22.5		

axis line. Instead of a least squares fit (summing squares of vertical distances from points to a line to a minimum—AB in figure 5.8) this line sums the areas of the triangle between points and the best-fit line to a minimum (CDE in figure 5.8). A unique line of best-fit is thus produced, as plotted in figure 5.9.

For a reduced major axis line (also referred to as the unique line of organic correlation or the isogonic growth line)

$$y = a + bx$$

The slope is

$$b = s_y/s_x$$

where s_y is the standard deviation of y and s_x the standard deviation of x. This reduces to

$$b = \left[\frac{\text{CSSY}}{\text{CSSX}}\right]^{\frac{1}{2}}$$

The value of the slope must be given the sign of the correlation coefficient, r. The intercept is

$$a = \bar{y} - b\bar{x}$$

These formulae can be derived in a similar manner to those demonstrated in the previous section.

For our data in table 5.4

$$n = 40$$

$$\sum x = 883 \qquad \sum x^2 = 20294.5$$

$$\sum y = 832 \qquad \sum y^2 = 18067.5$$

$$\sum xy = 19130.3$$

$$\bar{x} = 22.075, \qquad \bar{y} = 20.8$$

$$\text{CSCP} = 755.425$$

$$\text{CSSX} = 802.275$$

$$\text{CSSY} = 761.9$$

giving

$$r = 0.9770$$

$$b = \left[\frac{761.9}{902.275}\right]^{\frac{1}{2}} = 0.9745$$

$$a = 20.8 - 0.9745 \times 22.075 = -0.7124$$

and a reduced major axis line

$$y = -0.71 + 0.97\,x \tag{5.10}$$

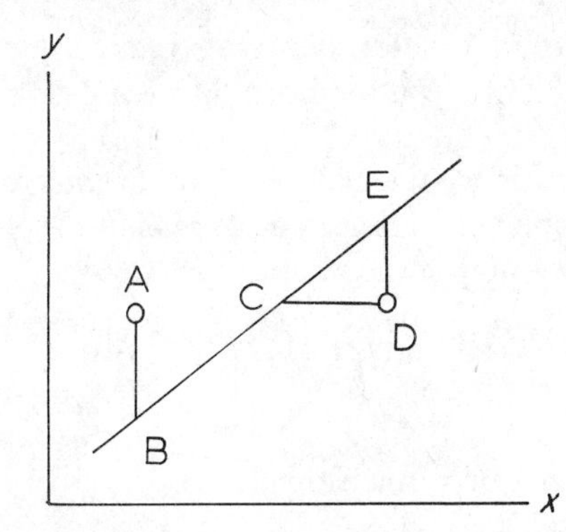

Figure 5.8 A comparison between the calculation of the classical regression and the reduced major axis line.

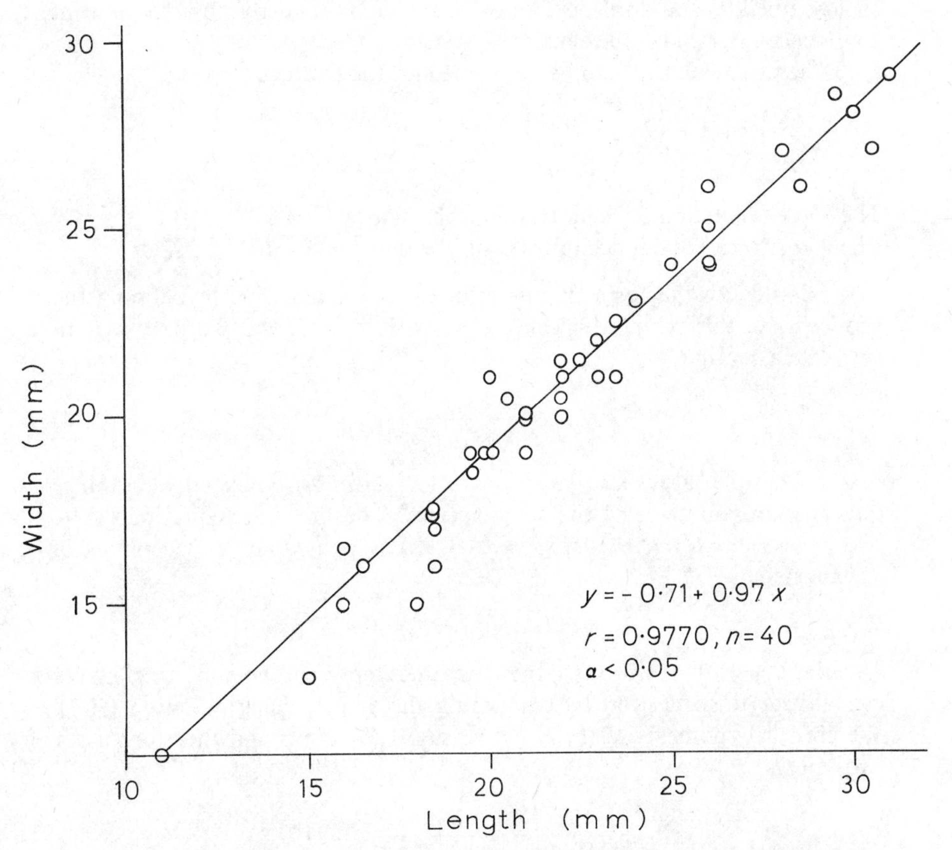

Figure 5.9 The reduced major axis line for the Nucleolites scutatus *measurements from table 5.4.*

Though many decimal places should be retained during the calculation, it is not realistic finally to quote more significant figures than the original data contained.

Again we may want to calculate confidence intervals on the slope and intercept of our sample line. This can be done in terms of standard deviations.

The standard deviation of the intercept is

$$s_a = s_y \sqrt{\left(\frac{1 - r^2}{n}\left[1 + \frac{\bar{x}^2}{s_x^2}\right]\right)}$$

The standard deviation of the slope is

$$s_b = b\sqrt{\left(\frac{1 - r^2}{n}\right)}$$

In the normal distribution 95 per cent of the area under the curve is enclosed in 1.96 standard deviations (see table 3.6). Therefore a 95 per cent confidence interval on the slope would be $b \pm 1.96\, s_b$.

The same calculation can be performed for the intercept, giving

$$s_b = 0.0329 \qquad 1.96\, s_b = 0.0644$$

$$s_a = 0.7404 \qquad 1.96\, s_a = 1.4512$$

The 95 per cent confidence interval on the slope $= 0.97 \pm 0.06$
The 95 per cent confidence interval on the intercept $= 0.71 \pm 1.45$

We could use the form of this reduced major axis line in defining the properties of this echinoid species. If we wish to compare two lines we can calculate a z-value

$$z = \frac{b_1 - b_2}{\sqrt{(s_{b_1}^2 + s_{b_2}^2)}} \tag{5.11}$$

For example, are some specimens of *Nucleolites* collected at another quarry representatives of the same species? The first line, equation (5.10), had parameters $b_1 = 0.9745$, $s_{b_1} = 0.0329$. Let us assume the second set of specimens has

$$b_2 = 1.0415 \qquad s_{b_2} = 0.0453$$

for a line $y = 1.31 + 1.04x$. Do these two lines represent the same species? We can answer this question by comparing the slopes, using equation (5.11). We erect a hypothesis $H_0(b_1 = b_2)$ versus $H_a(b_1 \neq b_2)$ and therefore use a two-tail test

$$z = \frac{0.9745 - 1.0415}{[(0.0329)^2 + (0.0453)^2]^{\frac{1}{2}}} = 1.197$$

From our tables of the area under a normal distribution (table 3.6) we find that a larger value of z could occur 23 per cent of the time if both slopes belong to the same population. Therefore we must accept the hypothesis that the two groups of echinoids are not significantly different in terms of the slope of their reduced major axis lines, and therefore probably belong to the same species.

Data transformations

Inspection of the (x, y) plot, or scatter diagram, may suggest that a straight line is not the best fit to a set of data (for example some length–breadth relationships in fossils). Often we can obtain a straight line by replotting the data on log–log graph paper. In the same way we can try and fit a straight line to the logarithms of the x–y values. For fossils this is called the allometric growth line. We now have

$$\log y = \log a + b \log x$$

(that is, $y = ax^b$) which we can use and test as above.

More will be said about transformations of data, and their ability to normalise skewed distributions, in the following chapter on variance analysis. Good discussions of regression methods can be found in Draper and Smith (1966) and Acton (1966).

5.4 Applications

Generation of a correlation matrix is often the first step in a more complex, multivariate statistical procedure. Examples of this can be seen in Vistelius *et al.*(1970), and Till and Colley (1973) where it forms the basis of a principal component analysis (see chapter 8). Macdonald and Bailey (1973) list a large set of lava whole-rock analyses and uses a correlation matrix to assess which chemical elements covary.

The chemical data for some recent carbonates from Billings and Ragland (1969) is treated by correlation and regression methods in Till (1971). Doornkamp and King (1971) give data sets and calculate classical regressions for many geomorphological features.

Taylor (1973) uses the reduced major axis line to calculate shear strength parameters for triaxial tests on colliery spoil. Till (1973) shows how and when we should use a reduced major axis line in geomorphology to define valley shapes and Parkinson (1954) gives an early example of its use in palaeontology.

6 ANALYSIS OF VARIANCE

We have used the t-test in chapter 4 to compare the means of two samples, but we often need to compare the mean values of larger numbers of samples. For example is there a significant difference between the mean diameter of four samples of a species of Jurassic pollen, is the porosity of five sandstones different? Such questions are often asked in all parts of the earth sciences and indeed in many other sciences as well. A comprehensive and powerful set of procedures, all based on the F-test devised by Fisher, originally for agricultural use, and referred to generally as 'analysis of variance' have been developed.

We shall consider two simple examples of analysis of variance, but there are many complex experimental designs for which it can be used. Krumbein and Graybill (1965), Griffiths (1967) and Koch and Link (1971) give full

Table 6.1 Salinity values (parts per thousand) for the three separate water masses in the Bimini Lagoon, Bahamas.

	Water mass 1	Water mass 2	Water mass 3
	37.54	40.17	39.04
	37.01	40.80	39.21
	36.71	39.76	39.05
	37.03	39.70	38.24
	37.32	40.79	38.53
	37.01	40.44	38.71
	37.03	39.79	38.89
	37.70	39.38	38.66
	37.36		38.51
	36.75		40.08
	37.45		
	38.85		
$n =$	12	8	10
$\sum x =$	447.76	320.83	388.92
$\bar{x} =$	37.31	40.10	38.89
$s^2 =$	0.3282	0.2826	0.2349

descriptions with geological examples. The books by Fryer (1966) and Li (1964) also describe and derive the methods fully, using non-geological examples.

Table 6.1 gives a set of salinity values for three separate water masses in a Bahamian lagoon. This data will be used as an example through the first analysis of variance procedure.

6.1 Homogeneity of variance

If we have k groups of data whose mean values we wish to compare, we erect a hypothesis, H_0, that the sample means we have represent population means that are equal

$$H_0(\mu_1 = \mu_2 = \mu_3 \ldots = \mu_k)$$

that is, for our data in table 6.1 H_0 asks: 'Is there a significant difference in salinity for the three water masses?' But this hypothesis assumes that all the population variances are equal

$$H_0(\mu_1 = \mu_2 = \mu_3 \ldots = \mu_k | \sigma_1^2 = \sigma_2^2 = \sigma_3^2 \ldots = \sigma_k^2)$$

This assumption is important, because if it is not met, the power of the test is reduced. Often people do not bother to test this assumption in making their analysis of variance. This is unfortunate, because if it is grossly untrue, a non-parametric test (see chapter 7) should be used in place of an F-test. Two simple tests are available to test the equality, or homogeneity, of population variances, and we should always use them before starting our F-test. The first test—called Hartley's maximum-F test—is a shorter and easier version of Bartlett's test for homogeneity of variance. Fryer (1966, section 7.2) gives full details of Bartlett's test which follows the χ^2 distribution.

Hartley's maximum-F test

Hartley's test explores the hypothesis H_0 that the variances of all the populations from which we have samples are the same

$$H_0(\sigma_1^2 = \sigma_2^2 = \sigma_3^2 \ldots = \sigma_k^2 = \sigma^2)$$

versus

$$H_a \text{ (some } \sigma^2 \text{ are not equal)}$$

We calculate

$$F_{max} = (\text{largest } s_i^2)/(\text{smallest } s_i^2) \tag{6.1}$$

where s_i is the sample estimate of σ_i, referred to as the sample standard deviation (or sample mean square).

Strictly, all samples should be of the same size and F_{max} has ν degrees of freedom, for $v = n - 1$, where n is the number of objects in each sample. Hartley shows that we can use an average value for v where the k groups are of equal size

$$v = \sum_{i=1}^{k} n_i/k$$

For our data in table 6.1

$$F_{max} = 1.40 \qquad \text{with 10 d.f. and } k = 3$$

Significant values of F_{max} can be interpolated from figure 6.1. The diagram plots F_{max} against v for a variety of k values. For 10 d.f. and $k = 3$, $F_{max} = 4.85$, which is much larger than our calculated value. Therefore we can accept the hypothesis H_0 that the variances are not significantly different. We can now proceed to apply an F-test, knowing that the data conforms to the assumptions required by the test.

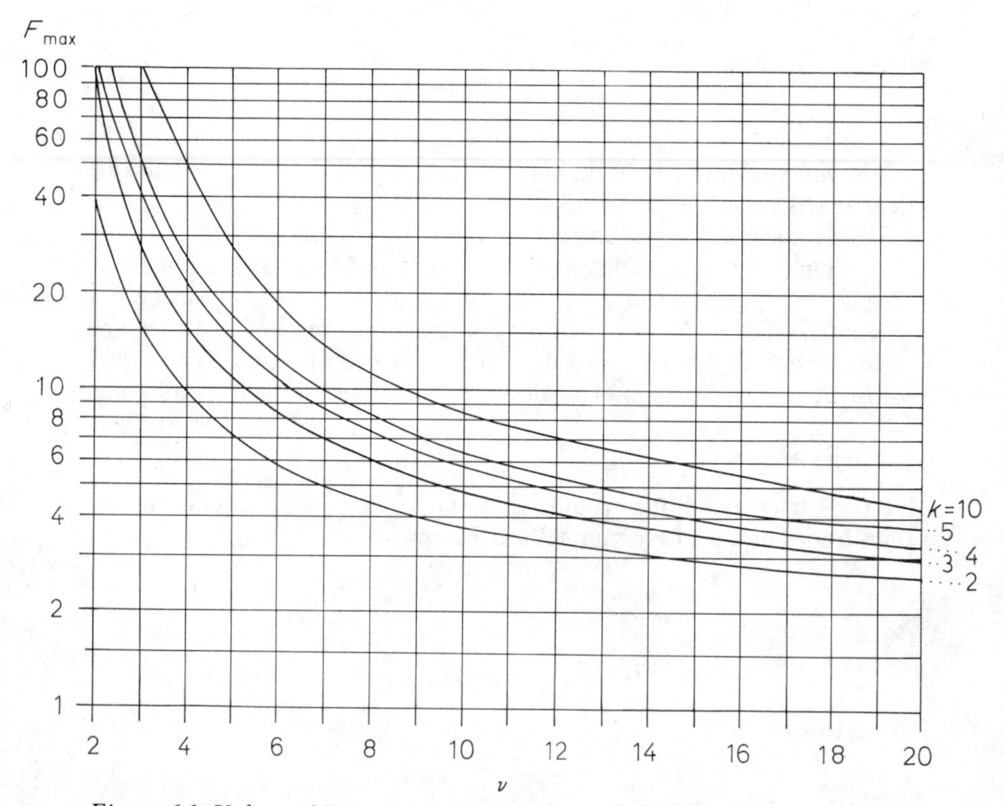

Figure 6.1 Values of F_{max} for various degrees of freedom v, and numbers of sample groups k, at the 5 per cent significance level. Produced from the data in Pearson and Hartley (1969, table 31).

Cochran's test

The purpose of this test is slightly different from that of Hartley's test. A serious deviation from homogeneity of variances is caused by one population variance being much larger than the others. This greatly affects the power of any variance analysis. Cochran's test is designed to explore this situation.

The null hypothesis in this test is

$$H_0(\sigma_1^2 = \sigma_2^2 = \sigma_3^2 \ldots = \sigma_k^2 = \sigma^2)$$

versus

$$H_a(\sigma_i^2 > \text{rest of } \sigma^2\text{s})$$

which is tested by

$$C = \frac{(\text{largest } s_i^2)}{\sum_{i=1}^{k} s_i^2} \tag{6.2}$$

Significant values of C can be interpolated from figure 6.2, using the procedure employed in figure 6.1. Each of the k samples should be of the same size, giving v degrees of freedom where $v = n - 1$ as before.

A value of C must be greater than the graphed value for the null hypothesis to be rejected.

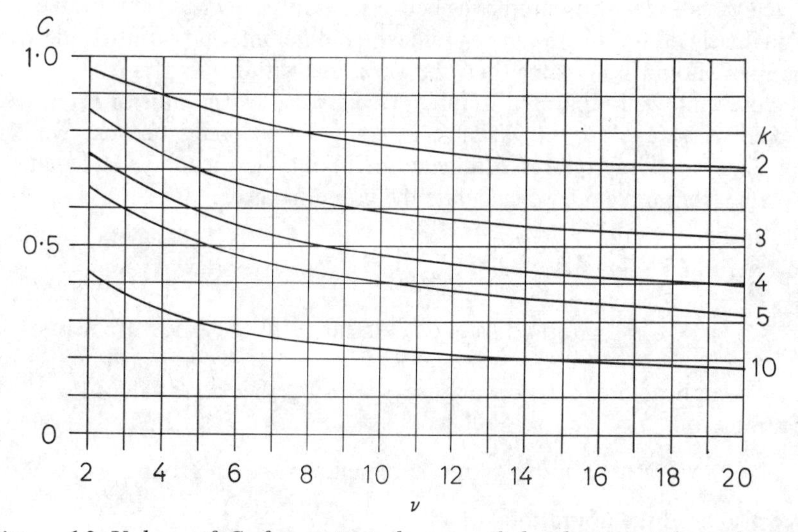

Figure 6.2 Values of C for various degrees of freedom, v, and numbers of sample groups, k, at the 5 per cent significance level. Produced from the data in Pearson and Hartley (1969, table 31a).

6.2 A simple (one-way) analysis of variance

Having established homogeneity of variance for our samples we can now proceed to ask the first question again: 'For the data in table 6.1 is there a significant difference in sea-water salinity in the three areas of the lagoon?' that is, 'Is there a significant difference between the population means which our samples represent?'

We erect a null hypothesis

$$H_0(\mu_1 = \mu_2 = \mu_3 | \sigma_1^2 = \sigma_2^2 = \sigma_3^2 = \sigma^2)$$

versus

$$H_a \text{ (some } \mu\text{s are not equal)}$$

Our use of analysis of variance depends on a simple algebraic identity which states that

[Total variance of a system] = [Within-samples variance]

+ [Between-samples variance]

The total variance is the variation of all measurements about their ground mean. The within-samples variance is the total variation of each sample about its own mean, and the between-samples variance is the variation of the sample means about the grand mean. Our F-test looks at the ratio of between-samples variance to within-samples variance. Intuitively we can see that if a number of means are not significantly different there will be as much variation within samples as between samples, giving a small value of F. Conversely if the means are significantly different the variation between samples should be greater than the variation within samples (that is, a set of close-knit, well-separated groups). This gives a larger value of F.

The name analysis of variance is a slight misnomer, because we only calculate sums of squares of deviations about the mean. To calculate an F-value from table 6.1 we calculate the grand mean $x_{..}$

$$x_{..} = \sum_{j=1}^{k} \sum_{i=1}^{n_j} x_{ij} \Big/ \sum_{j=1}^{k} n_k$$

$$= \text{(grand total of } x\text{)/(total number of measurements)}$$

This complicated expression only says add together all x values for all the k samples

$$n_j = \text{number of measurements in the } j\text{th group}$$

We also need the grand total of x^2

$$\text{(grand total of } x^2) = \sum_{j=1}^{k} \sum_{i=1}^{n_j} x_{ij}^2$$

For our data

$$n_1 = 12 \qquad n_2 = 8 \qquad n_3 = 10$$

Total number of measurements $= 30 = \sum n_k$

$$\text{grand total of } x = 1157.51$$

$$\text{grand total of } x^2 = 44705.7149$$

$$x.. = 1157.51/30 = 38.58$$

The between-groups variance is called the Between-Groups Sum of Squares (BGSS)

$$\text{BGSS} = \frac{(\text{group 1 sum})^2}{n_1} + \frac{(\text{group 2 sum})^2}{n_2} + \frac{(\text{group 3 sum})^2}{n_3}$$

$$\ldots + \frac{(\text{group } k \text{ sum})^2}{n_k} - \frac{(\text{grand total of } x)^2}{\sum n_k}$$

which is equivalent to, but easier to calculate than actual deviations of sample means from the grand mean

$$\text{BGSS} = \frac{(447.76)^2}{12} + \frac{(320.83)^2}{8} + \frac{(388.92)^2}{10} - \frac{(1157.51)^2}{30}$$

$$= 38.798$$

The total variance of the system is represented by the total sum of squares about $x..$(TSS)

$$\text{TSS} = (\text{grand total of } x^2) - \frac{(\text{grand total of } x)^2}{\text{total number of measurements}}$$

$$\text{TSS} = 44705.7149 - \frac{(1157.51)^2}{30} = 44.735$$

This leaves the Within-Groups Sum of Squares (WGSS)

$$\text{WGSS} = \text{TSS} - \text{BGSS}$$

$$= 44.735 - 38.798 = 5.937$$

We can now build our analysis of variance table, which is always presented in a standard format (table 6.2). The between-groups degree of freedom

$$v_1 = k - 1 = 2$$

the within-groups degrees of freedom

$$v_2 = \sum_{j=1}^{k} n_k - k = 12 + 8 + 10 - 3 = 27$$

and the total degrees of freedom about $x_{..}$ (which are not needed in the calculation)

$$= \sum_{j=1}^{k} n_k - 1 = 12 + 8 + 10 - 1 = 29$$

Dividing the sums of squares by their appropriate degrees of freedom gives us two estimates of variance in the samples (compare it with the calculation of s^2).

The between-groups Mean Square (also called the Treatment Mean Square)

$$= \text{BGSS}/v_1$$

the within-groups Mean Square (also called the Error Mean Square)

$$= \text{WGSS}/v_2$$

from there we calculate our F-test value:

$$F = \frac{\text{between-groups Mean Square}}{\text{within-groups Mean Square}} \tag{6.3}$$

From table 6.2 we calculate $F = 88.2$, which has v_1 and v_2 degrees of freedom. Using F-tables (for example, Murdoch and Barnes 1974, table 9) we find

$$F_{(0.05;2,27)} = 3.36$$

So we can reject H_0, since our value of F would arise far less than 5 out of a 100 times if the samples all belonged to the same population. We conclude that the salinity of all the water masses is not the same at a 5 per cent confidence level.

The calculation of F may seem a little confusing. It is worth following the whole procedure through with a few sets of data. The easiest way to understand it is to compare it with our familiar s^2. The Error Mean Square is really an s^2 for the sample means, and the within-groups Mean Square is a sum of s^2 for each sample group.

Table 6.2 Analysis of variance table for the data from table 6.1.

Source of variation	d.f.	Sum of squares	Mean square	F	Decision
Between-groups	2	38.798	19.399	88.2	Reject H_0
Within-groups	27	5.937	0.2199		
Total about $x_{..}$	29	44.735			

Fisher's least significant difference test

We have rejected a hypothesis H_0 that the sample means are all the same

$$H_0(\mu_1 = \mu_2 = \mu_3 | \sigma_1^2 = \sigma_2^2 = \sigma_3^2 = \sigma^2)$$

in favour of an alternative

$$H_a \text{ (some } \mu\text{s are not equal)}$$

We now ask: 'Which means are not equal?' We can answer this by using Fisher's least significant difference (LSD) test

$$\text{LSD}_\alpha = t\sqrt{\left(s^2\left\{\frac{1}{n_i} + \frac{1}{n_j}\right\}\right)} \tag{6.4}$$

where s^2 = Error Mean Square (from table 6.2)

t = α-level two-tail t-value for v_1 degrees of freedom (d.f. of s^2)

For this data $t_{(\alpha/2;27)} = 2.05$

n_i = size of sample i

n_j = size of sample j

This gives us an actual value of the measured variable (in this case salinity), by which groups i and j must differ for their means to be considered significantly different at the α-level of significance. The α-level used must be the same as that used for the F-test. Obviously it cannot be a higher level of significance since the means may not be different at that level, and there is no point in using a lower level.

For example, considering water masses 1 and 3 in table 6.1

$$\text{LSD}_{0.05} = 2.05\sqrt{(0.2199[1/12 + 1/10])}$$
$$\text{LSD}_{0.05} = 0.43$$

which tells us that the mean values of groups 1 and 3 are significantly different at the 5 per cent level. We calculate an LSD_α value for each pair of sample means and produce an ordered array of sample means

For groups 1 and 3: $\text{LSD}_{0.05} = 0.43$
For groups 1 and 2: $\text{LSD}_{0.05} = 0{\cdot}44$
For groups 3 and 2: $\text{LSD}_{0.05} = 0{\cdot}46$

Group (water mass) number	1	3	2
Mean salinity, $\bar{x}$	37.31	38.89	40.10

The LSD_α values show that all three means are significantly different at the 5 per cent level (we could have guessed this from the very large F-value we calculated). Consequently the salinities of each water mass are significantly different from one another and we have three distinct populations.

When this salinity data was collected, the underlying sediment was also studied. The content of skeletal material in each sediment specimen was one of the parameters recorded. An F-test on these observations showed that the content of skeletal material in the three areas of the lagoon was also significantly different (a reflection of the different aqueous environments?)

for group 1	$\bar{x} = 34.5\%$ skeletal matter
for group 2	$\bar{x} = 23.0\%$ skeletal matter
for group 3	$\bar{x} = 37.5\%$ skeletal matter

$F = 3.45$ with 2 and 27 d.f. which is just significant at the 95 per cent level. Calculation of LSD values gave

for groups 1 and 2	$\text{LSD}_{0.05} = 12.4$
for groups 1 and 3	$\text{LSD}_{0.05} = 12.0$
for groups 2 and 3	$\text{LSD}_{0.05} = 12.6$

From the group sample means it is obvious that groups 2 and 3 are significantly different, but groups 1 and 3 and groups 1 and 2 are not significantly different. We can make a tabular description of this situation using an ordered array of sample means

Group number (water mass)	2	1	3
Mean skeletal content, $\bar{x}$	23.0	34.5	37.5

Mean values that are not significantly different are joined by a line. Hence we see that groups 2 and 3 are the only significantly different populations.

Assumptions and transformations

Because of our definition of the F-test value in equation (6.3) and the theoretical generation of an F-value, a number of assumptions are implicit in the use of an F-test. The main assumption is that the population means are equal, but this is the condition being tested. The other requirements are

1. the k samples are random samples;
2. the k populations are normal;
3. the variance of the k populations are equal.

The requirement of randomness of the samples is most important—sampling procedures (see chapter 4) can be developed to ensure this condition is met. Slight non-normality of the populations does not seriously affect an F-test (or a t-test), nor does slight inhomogeneity of variances, but it is still worth testing these properties before beginning an analysis of variance. The effect of inhomogeneity can be reduced by making the size of all k samples the same.

Though statisticians have shown that slight departures from the assumptions cause no worries, large departures will seriously affect the power of our F-test. While recognising analysis of variance as a 'rugged test' we should not become too complacent about its use. If our data is markedly non-normal we can either use some non-parametric test (see next chapter) or else use a transformation on our data to make its distribution closer to normality. Several transformations are commonly used which involve replacing each measurement (x_i) in all k samples by some algebraic transformation, y_i, and then performing the analysis of variance on the y values.

The transformations are

$$y_i = \sqrt{(x_i + a)} \qquad \text{square-root transformation}$$

or

$$y_i = \ln(x_i + a) \qquad \text{logarithmic transformation}$$

or

$$y_i = \arcsin (x_i) \qquad \text{arcsin transformation}$$

where a is some constant.

The arcsin transformation is used for percentage data, and here x_i is the actual percentage value recorded. The square-root transformation can be used for data having a Poisson distribution. The choice of an appropriate transformation is a matter of experience; it is an empirical decision. However some distributions show a relationship between mean and variances which the normal distribution does not have (for example variance increases as the mean increases for some counting procedures). A graph of sample means versus variance before and after transformation should ideally show the removal of any relationship. With this suitable transformation the analysis of variance can then proceed. Fryer (1966, section 9.4) gives further details and examples of the application of transformations and how to deal with incomplete data sets.

6.3 A two-way analysis of variance

In the simple one-way analysis of variance we tested a single hypothesis H_0, about the mean value of each sample. To give an idea of how the complexities of variance analysis develop we can go one stage further, using the data in table 6.3. This shows three different geochemical techniques being applied to six different sediment specimens (or in accepted terminology, three treatments performed on six replicates). We can answer two questions with this data. Firstly, is there a significant difference between the six sediment specimens and secondly, is there a difference between the results obtained by different techniques? The second question is important to any analyst. If the results obtained by different techniques used on the same specimen are different, it is difficult to compare data sets. This problem becomes even more acute when we try to compare analytical results, obtained by one method, with those of another worker, obtained by some other method. This problem of comparing data sets plagues all fields of science and especially the earth sciences, where standard rocks, soils or waters for analysis are difficult to produce.

Since we look at the data in two ways (by rows and by columns) we refer to a two-way analysis of variance, which is applied to blocked data.

We can erect and test two null hypotheses, H_0 and H'_0

H_0 (all specimens have the same strontium content) versus
H_a (some specimens have different strontium contents) and

H'_0 (all methods give the same strontium values) versus
H'_a (some methods give different strontium values).

Computation is similar to the previous analysis of variance. We calculate the specimen totals, the method totals and the grand total as shown in table 6.3, together with the grand total of x^2.

Table 6.3 Strontium content (per cent Sr) of a set of carbonate sediments, using three different analytical methods. I—flame photometry, II—spectrographic analysis, III—atomic absorption.

Sediment specimen No.	Method I	Method II	Method III	Specimen totals
1	0.96	0.94	0.98	2.88
2	0.96	0.98	1.01	2.95
3	0.85	0.87	0.86	2.58
4	0.86	0.84	0.90	2.60
5	0.86	0.87	0.89	2.62
6	0.89	0.93	0.92	2.74
Method totals	5.38	5.43	5.56	16.37

If there are r specimens and t methods

$$\text{grand total of } x^2 = \sum_{i=1}^{r} \sum_{j=1}^{t} x_{ij}^2 = 14.9339$$

Total number of observations $= r \times t = 18$, so total sum of squares is

$$\text{TSS} = (\text{grand total of } x^2) - \frac{(\text{grand total})^2}{\text{total number of observations}}$$

$$= 0.046295$$

'Specimens' sum of squares is

$$\text{SS1} = \frac{\sum_{i=1}^{r} (\text{specimen total } i)^2}{t} - \frac{\sum_{i=1}^{r} (\text{specimen total } i)^2}{\text{total no. of observations}}$$

$$= \frac{(2.68)^2 + (2.95)^2 + (2.58)^2 + (2.60)^2 + (2.62)^2 + (2.74)^2}{3} - \frac{(16.37)^2}{18}$$

$$= 0.040828$$

Similarly, 'methods' sum of squares is

$$\text{SS2} = \frac{\sum_{j=1}^{t} (\text{method total } i)^2}{r} - \frac{\sum_{j=1}^{t} (\text{method total } i)^2}{\text{total no. of observations}}$$

$$= \frac{(5.38)^2 + (5.43)^2 + (5.56)^2}{6} - \frac{(16.37)^2}{18}$$

$$= 0.002878$$

which leaves the error sum of squares (ESS)

$$\text{ESS} = \text{TSS} - \text{SS1} - \text{SS2} = 0.002589$$

We obtain the appropriate Mean Squares by dividing a sum of squares by their degrees of freedom

For the 'specimens' sum of squares d.f. $= r - 1 = 5$
For the 'methods' sum of squares d.f. $= t - 1 = 2$
For the error sum of squares d.f. $= (t - 1)(r - 1) = 10$

In this way the variance has been partitioned into three sources; inter-method, inter-specimen and a remaining (normally distributed) random error.

We test H_0 by calculating an F-value

$$F = \frac{\text{Specimens Mean Square}}{\text{Error Mean Square}} \quad \text{with 5 and 10 d.f.}$$

and test H_0' by

$$F' = \frac{\text{Methods Mean Square}}{\text{Error Mean Square}} \quad \text{with 2 and 10 d.f.}$$

From tables

$$F_{(0.05;5,10)} = 3.33$$

and

$$F_{(0.05;2,10)} = 4.10$$

Table 6.4. Two-way analysis of variance.

Source of variation	d.f.	Sum of squares	Mean square	F	Decision
Specimens	5	0.040828	0.0081656	31.54	Reject H_0
Methods	2	0.002878	0.001934	7.47	Reject H'_0
Remainder (error)	10	0.002589	0.0002589		
Total about $x_{..}$	17	0.046295			

For our data (see table 6.4)

$$F = 31.54 \quad \text{and} \quad F' = 7.47$$

which means we can reject H_0 and reject H'_0 at a 5 per cent confidence level.

We conclude that there is a significant difference between the sediment specimens and also that the analytical techniques give significantly different results—rather worrying for the analysts involved!

6.4 Applications

Koch and Link (1970, chapter 5) give examples of several designs of variance analysis being used in estimating and comparing ore grades. Krumbein and Graybill (1965, chapter 9) also give a detailed description of analysis of variance models. Griffiths (1967, chapters 18–20) gives many examples from sedimentology. These include the use of F-tests in comparing different methods of porosity measurement and in comparing the lengths of quartzite boulders in screes.

Crozier (1973) uses F-tests to distinguish between different morphometric types of landslip. Hellewell and Myers (1973) use variance analysis to assess a geophysical technique, for *in situ* density measurements on rocks, which uses changes in gravity values. Carboniferous and Permo-Trias rocks in a borehole showed significant differences in the density recorded for different rock types (limestones, sandstones and marls).

7 NON-PARAMETRIC STATISTICS

The theory and use of non-parametric statistics is a large subject in itself. As the name suggests, non-parametric statistics use tests whose models do not specify conditions about the parameters of the population from which the sample was drawn. Such tests are usually applied to nominal or ordinal observations (see section 1.2)—the weaker levels of measurements.

The previous three chapters have been concerned with the parametric tests which assume

1. observations are independent;
2. the distribution of the population is known (usually normal);
3. the variances of populations being compared are equal or of known ratio,
4. measurements are at least on an interval scale.

Because of these clearly defined assumptions a properly used parametric test is very powerful. The non-parametric tests may in some cases be less powerful, but many types of data cannot fulfil some of these assumptions. This is the case in many parts of the earth sciences, where for example a simple observation of presence or absence of an object has been recorded. Also we may have sets of data which do not follow the normal distribution. For example, some palaeoecological measurements on communities produce skewed distributions. Some trace element data approximates to, but does not exactly follow, the log-normal distribution. In all these cases we can happily use some of the many non-parametric methods, which do not require us to specify the form of distribution which the data follows.

The following sections give an introduction to non-parametric tests. Further details can be found in the readable textbook by Siegel (1956). He describes the use of many tests for the behavioural sciences which are also applicable to geological data. Reyment (1971) gives examples of some non-parametric tests applied in quantitative palaeoecology.

7.1 Contingency tables (comparing sets of nominal data)

We are often faced with the problem of comparing two independent samples of observations to see if they indicate some differences in their parent populations. In parametric tests we used *t*-tests or variance analysis; here with measurements at the nominal scale, we set up a contingency table. For example, we may have noted the presence of heavy minerals in two suites of sandstones, or the presence/absence of some fossil on the bedding planes of two different Yoredale limestones. We may wonder if there is a significant difference between the sandstones, or between the limestones, in terms of the measured parameters. The data is tabulated in table 7.1.

Table 7.1 Relationships between sandstones and limestones in terms of their measured parameters.

	Tourmaline present	Apatite present	Total
Cretaceous sandstone	10	0	10
Jurassic sandstone	4	5	9
Total	14	5	19

	Fossil present	Fossil absent	Total
Limestone 1	6	3	9
Limestone 2	4	7	11
Total	10	10	20

The observations are nominal and the arithmetic only involves counting the number of Cretaceous sandstones which contain tourmaline, or the number of limestone outcrops which expose a specified fossil. This form of tabulation is called a contingency table and is displayed conventionally in table 7.2.

Table 7.2 A contingency table.

	+	–	Total
Group 1	a	b	$a + b$
Group 2	c	d	$c + d$
Total	$a + c$	$b + d$	n

Depending on the size of our sample, we should choose one of the following tests to compare our groups. The rules about which test to use are given at the end of the section.

Fisher's exact probability test

This can be used for nominal or ordinal data and should certainly be used if our two samples contain less than twenty objects in total ($n < 20$).

If we wish to compare groups 1 and 2 in table 7.2 for some observation we should erect a hypothesis H_0, that the noted distribution of observations (frequencies) arose by chance. We propose that there is no significant difference between group 1 and group 2. Fisher's test says that, given the marginal totals of table 7.2 are fixed (that is, our number of specimens in each group is fixed)—what is the probability, p, of our partitioning of frequencies (a, b, c, d) arising by chance

$$p = \frac{\binom{a+c}{a}\binom{b+d}{b}}{\binom{n}{a+b}}$$

where $\binom{a+c}{a}$ is a binomial coefficient $\dfrac{(a+c)!}{a!\,c!}$

$$p = \frac{(a+b)!\,(c+d)!\,(a+c)!\,(b+d)!}{n!\,a!\,b!\,c!\,d!} \tag{7.1}$$

An example will clarify the use of this test. Suppose we record the presence or absence of a particular living foraminifera in two different sedimentary environments as given in table 7.3.

Table 7.3 Presence/absence of living specimens of Cibicides lobatulus *in the sediment from around the islands and in the fine sand facies. Recent temperate carbonates from Mannin Bay, Connemara, W. Ireland. Data based on Lees* et al. (*1969, appendix 4*).

	Present	Absent	Totals
Islands fauna	6	6	12
Fine sand facies fauna	7	0	7
Total	13	6	19

We suspect that this species is found preferentially in the fine sand facies. Does our data uphold this suspicion?

We erect a null hypothesis H_0 that there is no difference between the two environments—saying the frequencies in our contingency table arose by chance. From equation (7.1)

$$p = \frac{12!\;7!\;13!\;6!}{19!\,6!\,6!\,7!\,0!} = 0.034$$

(note: $0! = 1$)

(It is well worth having statistical tables which include a good table of factorials, or else generating one by a simple computer program.)

This value of p tells us that, with our given numbers of samples the probability of the observed frequencies arising by chance is only 3.4 in 100. We should happily reject H_0, concluding that there is a significant difference between the two sedimentary environments, in terms of occurrence of the specified foraminifera, at the 5 per cent level.

This case is easy since one of the frequency cells (d) of the contingency table contains a zero, but if all cell frequencies are greater than zero we must calculate further values of p. This is because we must obtain the probability under H_0 of our case or a more extreme case (with the given marginal totals). Imagine the data was

	+	−	Total
I	5	2	7
II	3	6	9
Total	8	8	16

We are asking how likely is our partitioning of frequencies to arise by chance? This probability must include the whole 'extreme tail' of probability values—to represent our observations plus any more extreme situation. Really we are calculating what fraction of all possible partitionings that can occur is represented by our sample. So we compute

	+	−	T
I	5	2	7
II	3	6	9
T	8	8	16

$$p = \frac{7!\,9!\,8!\,8!}{16!\,5!\,2!\,3!\,6!} = 0.1371$$

	+	−	T
I	6	1	7
II	2	7	9
T	8	8	16

$$p = \frac{7!\,9!\,8!\,8!}{16!\,6!\,1!\,2!\,7!} = 0.0196$$

	+	−	T
I	7	0	7
II	1	8	9
T	8	8	16

$$p = \frac{7!\,9!\,8!\,8!}{16!\,7!\,0!\,1!\,8!} = 0.0007$$

$$p = 0.1371 + 0.0196 + 0.0007 = 0.1574$$

and we conclude that there is no significant difference at the 5 per cent level between the two sedimentary environments.

This test is very useful for treating small sets of data, but it becomes very tedious to compute when the smallest cell frequencies depart far from zero. In addition the factorials of quite small numbers soon become very large (for example, $20! = 2.44 \times 10^{18}$), which stretches the facilities of most electronic calculators or computers. Fortunately, a second powerful test is available in these cases.

The chi-square test

Besides the goodness-of-fit tests of chapter 4, analysing contingency tables is the other common situation in which the χ^2 test is used. Again we are considering nominal data, but the contingency table can be larger than the previous 2×2 case. Table 7.4 gives some example data

Table 7.4 Frequency of occurrence of different types of tourmaline in two sandstones encountered in a suite of boreholes.

	Sandstone A	Sandstone B	Totals
Green tourmaline	10	15	25
Yellow tourmaline	20	7	27
Pink tourmaline	6	8	14
Totals	36	30	66

It is suspected that two sandstone bodies encountered in a number of boreholes can be distinguished by the presence of different types of the heavy mineral tourmaline. Table 7.4 gives the frequency of occurrence of the minerals. We erect a null hypothesis, H_0, that there is no significant difference between the mineral content of the sandstones. As with Fisher's test, this hypothesis says that, given the marginal totals, the observed frequencies could arise by chance. We test this hypothesis by calculating an expected frequency for each cell.

For example, green tourmaline was observed in 10 samples of sandstone A. By chance we would expect 25/66 of samples to contain green tourmaline (since out of 66 specimens 25 contained tourmaline: marginal totals in table 7.4) and 36/66 to be sandstone A. Using our probability formulae from chapter 2 we expect a proportion of $(25/66) \times (36/66)$ of the samples to be sandstone A containing green tourmaline. To be rigorous, given the marginal totals, the probability of a green-tourmaline-sandstone-A is $(25/66) \times (36/66)$. In 66 samples we should get

$$\frac{25}{66} \times \frac{36}{66} \times 66 = \frac{25 \times 36}{66} \text{ such samples}$$

This last figure is our expected frequency.

To test our hypothesis H_0 we calculate

$$X^2 = \sum_{i=1}^{r} \sum_{j=1}^{k} \frac{(O_{ij} - E_{ij})^2}{E_{ij}} \tag{7.2}$$

which follows χ^2 with $(r - 1) \times (k - 1)$ degrees of freedom, where

O_{ij} is the observed value in the ith row of the jth column of our contingency table

and

E_{ij} is the expected value in the ith row of the jth column of our contingency table

($\sum\sum$ means sum values over all rows and all columns—that is sum the expression for all cells of the contingency table)

$$E_{ij} = \frac{T_i \times T_j}{n}$$

where

T_i is marginal total for row i
T_j is marginal total for column j
n is the total number of observations

$$\left(n = \sum_{i=1}^{r} T_i = \sum_{j=1}^{k} T_j\right)$$

and there are r rows and k columns in the contingency tables (compare this chi-square with section 4.5).

Table 7.5 gives the observed and expected frequencies for our data from which we calculate X^2.

$$X^2 = \frac{(10 - 13.7)^2}{13.7} + \frac{(15 - 11.3)^2}{11.3} + \frac{(20 - 14.7)^2}{14.7}$$
$$+ \frac{(7 - 12.3)^2}{12.3} + \frac{(6 - 7.6)^2}{7.6} + \frac{(8 - 6.4)^2}{6.4}$$

with $(2 - 1) \times (3 - 1)$ d.f.

$$X^2 = 0.99 + 1.21 + 1.91 + 2.28 + 0.34 + 0.4$$
$$= 7.13 \text{ with 2 d.f.}$$

Using tables of chi-square values we find that our value occurs at $\alpha < 0.05$ ($\chi^2_{(0.05;\,2)} = 5.99$). So we would reject our H_0, saying that because such a distribution of frequencies would arise by chance less than 5 in 100 times, there is a significant difference in heavy mineral content of the two sediments.

The contingency table can have any number of rows and columns, provided that all frequencies are greater than five (with smaller frequencies the power of the test is reduced). Most commonly a 2×2 table is used (as with the Fisher's test), in which case equation (7.2) does not follow the chi-square distribution completely. We use another equation to calculate

Table 7.5 Observed and expected frequencies for our two sandstones. The expected frequency is recorded in the lower right-hand corner of each cell.

	Sandstone A		Sandstone B		Totals
Green tourmaline	10	13.7	15	11.3	25
Yellow tourmaline	20	14.7	7	12.3	27
Pink tourmaline	6	7.6	8	6.4	14
Totals	36		30		66

X^2, which contains a correction (Yates' correction for continuity) to make the formula follow the chi-square distribution more closely and is also easier to compute. Following the convention of table 7.2

$$X^2 = \frac{n(|ad - bc| - n/2)^2}{(a + b)(c + d)(a + c)(b + d)} \tag{7.3}$$

with 1 d.f.

Table 7.6 Frequency of occurrence of a species of non-marine lamellibranch at outcrops of two Coal Measure shales.

	Present	Absent	Total
Shale I	11	5	16
Shale II	8	16	24
Total	19	21	40

We can use this to test H_0 for the data in table 7.6. Is the lamellibranch equally likely to occure in both Coal Measure shales?

Using equation 7.3

$$X^2 = 40\,\frac{(|11 \times 16 - 5 \times 8| - 40/2)^2}{16 \times 24 \times 19 \times 21}$$

$$= 3{\cdot}51 \quad \text{with 1 d.f.}$$

$$\chi^2_{(0.05;\,1)} = 3.84$$

So we cannot reject H_0, and must accept that there is no significant difference between the shales.

Equation (7.3) is used on exactly the same contingency table as the Fisher's test (equation 7.1). How do we know which to use? Statisticians have explored the power of these tests and concluded for a 2×2 table that

1. If the total number of observations (n) is less than twenty, use Fisher's exact probability test.
2. If n is less than forty, but more than twenty, use chi-square if all expected frequencies (E_{ij} in equation 7.2) are greater than five. If not, use Fisher's test.
3. Where n is greater than forty, use the chi-squared test.

In contingency tables with more than one degree of freedom (that is, more than two rows and/or columns) Siegel (1956) states that chi-square tests may be used powerfully if fewer than 20 per cent of the cells have an expected frequency less than five and if no cell has an expected frequency of less than 1. If this requirement is not met, all categories must be combined to meet it (or else more data collected!) before X^2 will follow χ^2 closely.

7.2 Comparing two sets of ordinal data

The Wilcoxon–Mann–Whitney test

Three statisticians had a hand in developing this 'non-parametric *t*-test', though it is also referred to as the Mann–Whitney *U*-test in many text books. It requires data to be measured at least on the ordinal scale, but is often used as an alternative to a *t*-test when the parent population is not known to be a normal distribution.

This test, referred to as the WMW test, is used to search for central tendency (or location) differences between two samples that might represent different populations. An example is computed using the data in table 7.7. This gives semi-quantitative spectrographic analyses for two stratigraphically equivalent sections in the Kimmeridge clay. A rapid visual-comparison method was used to estimate p.p.m. vanadium in the clay. Is there a significant difference between the vanadium content of these two clays? We cannot use a *t*-test because the data is semi-quantitative, which means that it is really only on an ordinal scale.

We propose a hypothesis, H_0, that the two samples come from populations that are not significantly different. The WMW test is usually a one-tailed test—and was certainly developed as such. Our alternative hypothesis, H_a, will be that the vanadium content of the clays in Dorset (referred to as $D(x_1)$, where $D(x)$ means the total distribution of values) is greater than its content at Boulonnais ($D(x_2)$). That is

$$H_0(D(x_1) = D(x_2))$$

versus

$$H_a(D(x_1) > D(x_2))$$

Table 7.7 Content of vanadium (p.p.m.) in two short sections of the Lower Kimmeridge clay from the Boulonnais, NW France and Dorset, England. Results are semi-quantitative. Data provided by C. Dunn.

Dorset		Boulonnais	
170	130	80	105
150	50	95	100
90	120	55	150
110	105	80	50
100	180	80	70
105	120	80	70
125	80	50	100
90	145	155	75
100	160	85	35
70	60	100	50

The two samples are ranked in a combined array, but taking note of the source of each measurement

Value: **180**, **170**, **160**, 155, **150**, 150, **145** ...

Rank: **1**, **2**, **3**, 4, **5.5**, 5.5, **7** ...

Value: 70, **60**, 55, **50**, 50, 50, 50, 35

Rank: 33, **34**, 35, **37.5**, 37.5, 37.5, 37.5, 40

Where two or more specimens have the same value on average rank for the tied values it is calculated as shown in the combined array above.

Values from sample 1 are in **bold type**. We compute the statistic

$$U = n_1 n_2 + \frac{n_1(n_1 + 1)}{12} - T \tag{7.4}$$

where

n_1 = number of measurements in sample 1

n_2 = number of measurements in sample 2

T = sum of the ranks for sample 1

If our alternative hypothesis, H_a, is that $D(x_2) > D(x_1)$, T must be the sum of the ranks for sample 2

$$n_1 = 20$$

$$n_2 = 20$$

$$T = 311$$

From equation (7.4)

$$U = 400 + 35 - 311 = 124$$

We could prove mathematically that U is normally distributed with

$$\mu_U = n_1 n_2/2, \text{ and } \sigma_U^2 = n_1 n_2(n_1 + n_2 +)/12$$

provided that n_1 and $n_2 > 8$.

If we standardise U, we can then test H_0, by using our tables of $N(0, 1)$

$$\lambda = \frac{U - (n_1 n_2/2)}{\sqrt{\{n_1 n_2(n_1 + n_2 + 1)/12\}}} \quad \text{follows } N(0, 1) \qquad (7.5)$$

Therefore the 5 per cent region of rejection for H_0 of the one-tailed U-test is $|\lambda| > 1.645$ (table 3.6) for all samples of size greater than eight. If either n is less than eight, significant values of U can be found in Siegel (1958, table K).

For our data

$$|\lambda| = \frac{|124 - 200|}{\sqrt{1366.7}} = \frac{76}{36.968}$$

$$= 2.056$$

Therefore we reject H_0 at the 5 per cent level, concluding that there is a significant difference in vanadium content of the populations from which our two samples were taken.

Ties can occur in the ranking process above in two ways. Two or more measurements can have the same value either in one sample or between the two samples. The former has no effect on U, but the latter affects U and its standard deviation σ_U. A correction can be applied as described by Siegel (1956, p. 124–5) which tends to increase the value obtained for the test, making the conclusion more statistically significant. To remain conservative in our conclusions the correction for ties can be ignored for most cases. Only if many observations (greater than 75 per cent) are involved in ties and many of these ties involve more than five observations is it worth making the correction.

The Kolmogorov–Smirnov two-sample test

This test is used to compare the complete form of two distributions and will respond to any differences in their form (for example, location, skewness, dispersion). Again it requires at least ordinal data, but is also used for testing measurements on higher scales whose population distribution is unknown. This makes it useful in geological situations where normality cannot be established or is known not to apply.

For our example, table 7.8 gives data on fluorine content for some over-saturated peralkaline lavas. We suspect that the lava types from the two

Table 7.8 Fluorine content (per cent) of continental and oceanic peralkaline oversaturated lavas. Data provided by D. K. Bailey.

Continental				Oceanic	
0.324	0.410	0.291	0.310	0.230	0.237
0.300	0.300	0.129	0.360	0.160	0.286
0.636	0.465	0.410	0.352	0.180	0.140
0.610	0.420	0.455	0.257	0.340	0.174
0.623	0.702	0.170	0.300	0.290	0.190
0.752	0.440	0.360	0.329	0.260	0.180
0.638	0.415	0.340	0.356	0.085	0.180
0.755	0.563	0.360	0.318	0.327	0.325
0.635	0.409	0.295	0.356	0.335	0.316
0.549	0.739	0.310	0.353	0.294	0.157
0.647	0.737	0.303	0.098	0.250	0.238
0.617	0.780	0.316	0.106	0.404	0.189
0.400	0.348	0.439	0.260	0.164	

different regions have a significantly different fluorine content. Though this data is measured on a ratio scale we must use a non-parametric test, because the distribution of values (shown in figure 7.1) does not approximate to any standard form we have described. The hypothesis, H_0, is that there is no significant difference between the two distributions. We can use a one-tailed or a two-tailed test. For a one-tailed test the alternative hypothesis, H_a, is

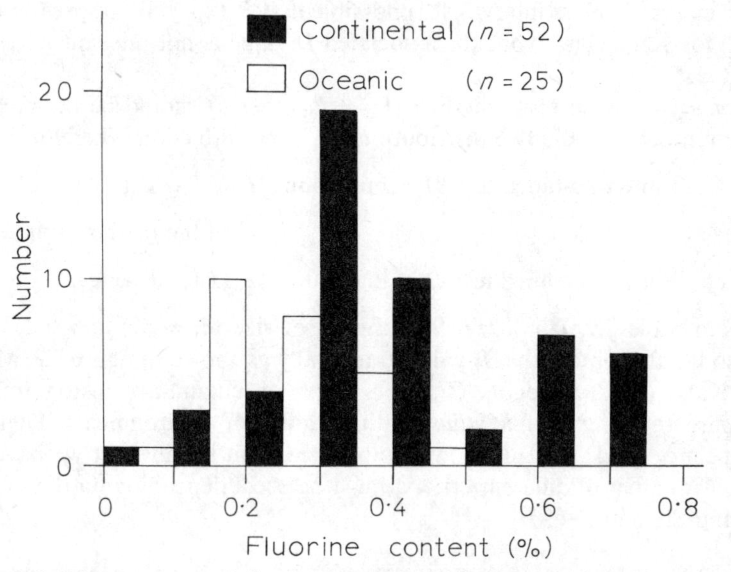

Figure 7.1 Histograms of fluorine content for continental and oceanic peralkaline oversaturated lavas, produced from the data in table 7.8. Class interval is 0.1 per cent for both groups of lavas.

that the first sample represents a population with a larger central value than does the second sample. For a two-tailed test, H_a is merely that the populations represented by the samples are different.

We suspect that the continental lavas (F_1) are richer in fluorine than the oceanic lavas (F_2), so we will choose a one-tailed test and $H_a(F_1 > F_2)$. The Kolmogorov–Smirnov test concentrates on the greatest difference between frequencies for a given class interval in the distributions. We produce cumulative relative frequency distributions for both samples, using the same class intervals. These are given in table 7.9 for the fluorine data. By inspection we find the largest difference in frequency, for a given class, between the two samples. Written formally

$$D = \text{maximum}\ \{F_1(x_i) - F_2(x_i)\} \tag{7.6}$$

where

F_1 is the frequency distribution of the first sample

F_2 is the frequency distribution of the second sample

i is the class for which D is a maximum

Note that for a one-tailed test, D is the maximum value of D in the correct predicted direction. If we think $F_1 > F_2$, then in figure 7.2 the correct D is the maximum separation upwards from F_1 to F_2. Remember that for the numerically larger distribution the cumulative relative frequency should be lower in a given class interval. It is possible that $F_2(x_i)$ could be greater than $F_1(x_i)$ for some class i, but the associated D-value is not relevant to a one-tailed test.

For a two-tailed test, where $H_a(F_1 \neq F_2)$ then D should be the greatest deviation between the two distributions, in either direction. Therefore

For a one-tailed test $D = \text{maximum}\ \{F_1(x_i) - F_2(x_i)\}$

in the predicted direction

For a two-tailed test $D = \text{maximum}\ \{F_1(x_i) - F_2(x_i)\}$

From table 7.9, $D = 0.606$ for samples of size ($n_1 = 52$) and ($n_2 = 25$). We could also obtain this D-value graphically as shown in figure 7.2. Miller and Kahn (1962, appendix C) use a graph of 'cumulative histograms' to compare the length of *Mytilus* in three different environments. Defining significance levels for D is more complicated than for any test we have yet used. A number of different criteria must be used, depending on the sample size and the nature of H_a.

For small samples where $n_1 = n_2$ and they are 40 or less. For this case n_1 must equal n_2 and critical values of D can be read off the step diagram in figure 7.3 for either a one-tail or a two-tail test.

Table 7.9 Cumulative relative frequency distributions for the data from table 7.8, showing the difference D, *in relative frequency, for each class interval.*

Fluorine content (%)	Cumulative relative frequency Continental	Oceanic	D
<0.1	0.019	0.04	0.021
0.1—0.2	0.077	0.44	0.363
0.2—0.3	0.154	0.76	0.606*
0.3—0.4	0.519	0.96	0.441
0.4—0.5	0.712	1.0	0.288
0.5—0.6	0.75	1.0	0.25
0.6—0.7	0.885	1.0	0.115
>0.7	1.0	1.0	0

* Maximum value of D.

For large samples where either n_1 or n_2 is larger than 40, and do not need to be equal.

For a one-tailed test statisticians have proved that

$$X^2 = 4D^2 \frac{n_1 n_2}{n_1 + n_2} \tag{7.7}$$

follows χ^2 with 2 d.f.

For a two-tailed test

$$D > 1.36 \left[\frac{n_1 + n_2}{n_1 n_2}\right]^{\frac{1}{2}} \tag{7.8}$$

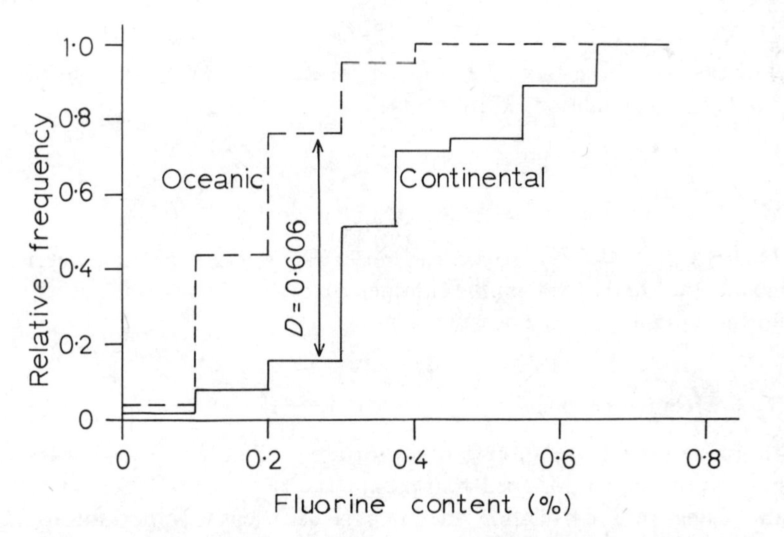

Figure 7.2 Cumulative relative histograms to determine the Kolmogorov–Smirnov D for the data from table 7.8.

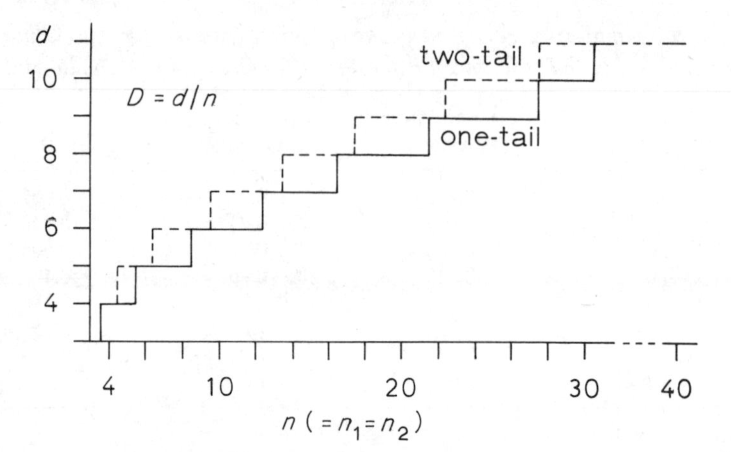

Figure 7.3 One-tailed and two-tailed test 5 per cent critical values of the Kolmogorov–Smirnov D for small samples ($n < 40$ and $n_1 = n_2 = n$). Graphed from data in Siegel (1956, table L).

allows us to reject H_0 at the 5 per cent significance level, and

$$D > 1.63\left[\frac{n_1 + n_2}{n_1 n_2}\right]^{\frac{1}{2}} \tag{7.9}$$

allow us to reject H_0 at the 1 per cent level. Appropriate values of D can be read off from figure 7.4 for values of N, where

$$N = \frac{n_1 + n_2}{n_1 n_2}$$

For the data from table 7.9, we calculated $D = 0.606$, with $n_1 = 52$ and $n_2 = 25$. Using equation (7.7)

$$X^2 = 4 \times (0.606)^2 \times \frac{52 \times 25}{77}$$
$$= 24.80 \quad \text{with 2 d.f.}$$

Now $\chi^2_{(0.05;2)} = 5{\cdot}991$, so we can reject H_0 at the 5 per cent level and conclude the alternative that the continental lavas have a significantly higher fluorine content than the oceanic lavas.

7.3 Correlation methods

Non-parametric correlation procedures are probably the most used and certainly the best developed non-parametric tests. Correlation of ranked data, using rank-correlation methods is sufficiently important to have merited a text book (Kendall, 1970). We shall only consider one such test, but Siegel (1956) and of course Kendall (ibid.) describe many other procedures.

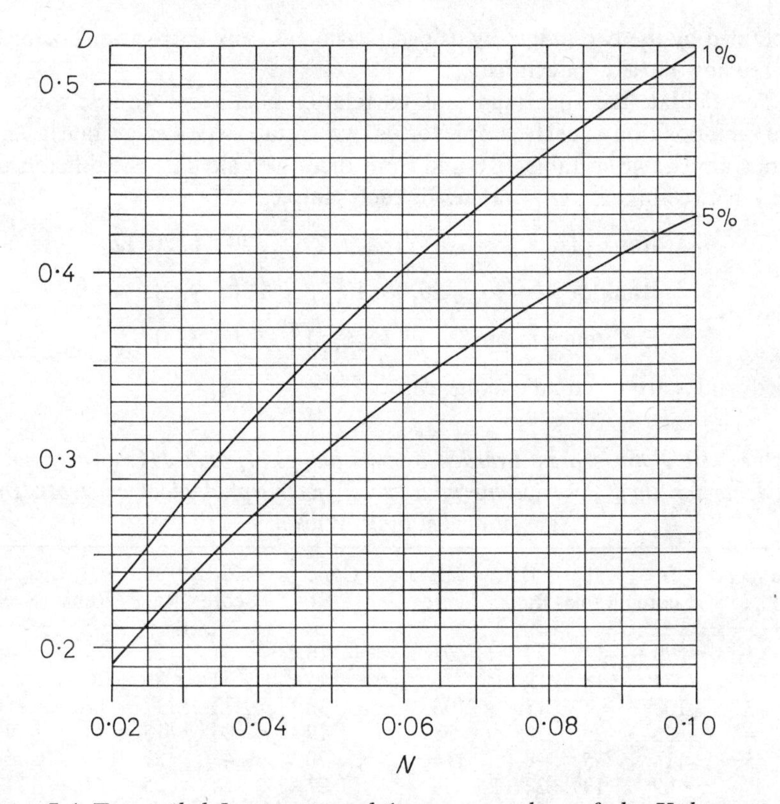

Figure 7.4 Two-tailed 5 per cent and 1 per cent values of the Kolmogorov–Smirnov D for large samples (n_1 or $n_2 \geqslant 40$) graphed versus N as defined in equations (7.8) and (7.9).

The Spearman rank correlation coefficient

This statistic may be used where only ranked data has been collected by necessity or for economy. Again it may be used where the product–moment correlation coefficient, r, is unsuitable because the population is not bivariate normal (see section 5.1). Table 7.10 gives the sets of scores awarded by two judges to a group of students for their descriptions of thirty concepts in geomorphology. Thus judge I gave a total of 79 points for students' descriptions of the first concept (landform) and 71 points for descriptions of the second concept (joints). Prior to a study of which concepts in geomorphology were best comprehended by the students, it was necessary to check that all the markers (judges) were giving comparable marks to the students. From table 7.10 we want to see if these two judges show a good correlation between the way in which they award marks for comprehension of a concept. Both judges marked the same answers. We can compare the rank order of concepts obtained by the two judges by using Spearman's rank correlation coefficient to see how closely they agree.

To calculate the Spearman rank correlation coefficient we first work out the rank position of each sample for its two variables (x and y values). These ranks are shown in table 7.10 and from them we calculate the difference in rank between the x and y values for each sample.

Rank of x	14, 18, 7, 29, ... 30, 6, 20, 12
Rank of y	20, 19, 13, 28, ... 30, 9, 21, 14
Difference in rank	6, 1, 6, 1, ... 0, 3, 1, 2

Tied values are given an average rank.

Table 7.10 Total scores awarded by two judges, JI and JII, to a group of students for their descriptions of a set of geomorphological concepts. Data provided by P. Wilson.

Concept No.	JI Score	JI Rank	JII Rank	JII Score	Concept No.	JI Score	JI Rank	JII Rank	JII Score
1	79	14	20	62	16	64	21	23	45
2	71	18	19	66	17	52	26	26.5	39
3	108	7	13	96	18	72	17	17	76
4	35	29	28	33	19	86	13	12	97
5	114	5	8	116	20	41	27	25	41
6	69	19	22	55	21	56	25	7	117
7	63	22	26.5	39	22	92	11	15	84
8	121	1	2	133	23	62	23	18	73
9	61	24	24	44	24	99	10	10	104
10	120	2	11	103	25	107	8	3	129
11	118	3	1	137	26	78	15	16	78
12	102	9	5	122	27	24	30	30	12
13	75	16	4	125	28	109	6	9	111
14	37	28	29	31	29	67	20	21	57
15	117	4	6	119	30	90	12	14	93

Our hypothesis, H_0, is that the correlation between x and y, that is their tendency to vary together, has arisen by chance. We calculate Spearman's coefficient

$$r_s = 1 - \frac{6 \sum_{i=1}^{n} d^2}{n(n^2 - 1)} \tag{7.10}$$

where

d is the difference between the x and y rank of an object

n is the number of pairs of observations

r_s can vary between $+1$ and -1.

For our data, $n = 30$ and

$$\sum d^2 = 36 + 1 + 36 + 1 \ldots + 0 + 9 + 1 + 4 = 785.5$$

giving

$$r_s = 1 - \frac{6 \times 785.5}{30(30 \times 30 - 1)}$$

$$= 0.8253$$

We could get a rough estimate of r_s by plotting a figure. One variable x is arranged in ascending rank order and the y rank for the same object is placed opposite to it. Lines are drawn joining the same rank in x and y. If $r_s = +1$, that is the two sets of ranks show a perfect positive correlation, then none of the lines joining ranks will cross. If the two sets of ranks show a perfect antipathetic relationship (that is, $r_s = -1$) then all lines will cross through the same point. All values of r_s will produce diagrams which lie somewhere between these two extremes.

How can we test the significance of r_s? Kendall (1970) proposed a rank correlation coefficient of his own, τ, because he maintains that r_s is difficult to test. However his τ is more difficult to compute, and most statisticians feel happy that the standard deviation of r_s can be computed because its $\sigma_{r_s} = \sqrt{\{1/(n-1)\}}$

From this we calculate

$$Y = r_s/\sigma_{r_s}$$

which is a standard normal deviate, $N(0, 1)$.

From equation (7.10), we found

$$r_s = 0.8253$$

$$\sigma_{r_s} = \sqrt{\{1/(30-1)\}} = 0.1875$$

and

$$Y = 0.8253/0.1857 = 4.44$$

Using table 3.6 this tells us that our value of r_s would arise by chance far less than once in a thousand times. Therefore we reject H_0 at the 5 per cent, and accept the alternative hypothesis that the two sets of ranks show a significance correlation. This is fortunate, and tells us that our two judges do agree.

If $n > 10$, a t-test can be used to test H_0

$$t = r_s\sqrt{\left(\frac{n-2}{1-r_s^2}\right)} \qquad \text{with } (n-2) \text{ d.f.}$$

This is a two-tailed t-test, since our alternative hypothesis is $H_a(r_s \neq 0)$. For the data above

$$t = 7.75 \text{ with } 28 \text{ d.f.}$$

which again lets us reject H_0 at the 5 per cent confidence level ($t_{(0.05/2;\,28)} = 2.05$).

A correction for ties in the x or y ranks can be made in the calculations for r_s (see Siegel, 1956, p. 207–9) but the effect is very small. Not correcting for tied ranks causes us to overestimate r_s slightly, but only in the third, or possibly the second significant figure.

Coefficients of similarity

Particularly in palaeoecology we may wish to compare objects for which we have only obtained data on a nominal scale. For example we may have recorded the presence or absence of a number of spore species in Devonian shales encountered in a suite of boreholes (table 7.11). The problem is to define similarity of spore assemblage between any two specimens.

Table 7.11 Occurrence of diagnostic spore species in Devonian shales from a suite of boreholes. Present is coded '2', absent '1' and no information '0'.

Species / Specimen	A	B	C	D	E	F
1	2	0	1	2	2	1
2	1	2	2	1	2	2
3	1	2	2	1	2	2
4	1	1	1	1	2	1
5	2	1	1	1	2	1

A number of coefficients of similarity have been defined, based on the presence of matches and mismatches between the specimens. Sokal and Sneath (1963) defined a simple matching coefficient, S_{SS}, where

$$S_{SS} = \frac{p + n}{p + n + m} \tag{7.11}$$

where

p = number of positive matches (defined as a character present in both specimens, that is, a matching of a '2' with a '2'),

n = number of negative matches (a character absent in both specimens—a '1' with a '1') and

m = number of mismatches (presence with absence or vice versa—'1' with '2' or '2' with '1').

For shales 1 and 2 in table 7.11

$$p = 1, \ n = 0, \ m = 4$$

All pairs of characters involving a '0'—an uncertainty—are ignored, giving

$$S_{SS} = \frac{1 + 0}{1 + 0 + 4} = 0.2$$

The similarity coefficient can very from zero, completely dissimilar, to unity, identical specimens.

For specimens 2 and 3

$$S_{SS} = \frac{4 + 2}{4 + 0 + 2} = 1$$

Another useful coefficient was defined by Jaccard and is called Jaccard's coefficient of association, S_J where

$$S_J = \frac{p}{p + m} \tag{7.12}$$

where symbols have the same meaning as in equation 7.11. This coefficient ignores negative matches. Only specimens with presences in common appear similar.

For specimens 4 and 5

$$S_{SS} = \frac{1 + 4}{1 + 4 + 1} = \frac{5}{6}$$

but

$$S_J = \frac{1}{1 + 1} = \frac{1}{2}$$

The choice of a suitable coefficient of similarity depends on the purpose in hand. Besides their use in palaeoecology, these coefficients can be used to compare two species with a set of defined characters. For example two acritachs might be compared for size (large/small), shape (round/not round), spines (present/absent), etc. The more of these characters two species have in common, the more similar they are. This type of procedure is the basis of numerical taxonomy, which is described fully in Sokal and Sneath (1963) and Sneath and Sokal (1973). Sokal and Sneath (1963) proposed to classify and recognise 'species' by a series of numerical criteria for which they erect an OTU (operational taxonomic unit). The members of an OTU could be defined by a certain level of similarity. Here the correct choice of similarity coefficient is important—for example the absence of wings (that is, a negative match) is not very useful in erecting a grouping within trilobites! But it is important in comparing trilobites with insects.

With the product moment correlation coefficient we made a matrix of all inter-variable correlation coefficients (see section 5.1) as the starting point of a number of multivariate procedures. We can use our similarity coefficients in a similar manner. Table 7.12 gives some petrographic data for Carboniferous limestone specimens. Instead of just using present (2) or absent (1),

Table 7.12 Coded presence/absence data for the petrographic constituents of a few Tournaisian carbonates from the Northumberland Basin. Data provided by M. Leeder.*

Character / Specimen	Crinoids	Brachiopods	Bivalves	Ostracods	Ooliths	Peloids	Coated grains	Algal pisoliths	Intraclasts	Quartz silt
1	111	111	221	221	221	211	221	111	211	211
2	211	111	222	221	111	221	111	211	211	211
3	111	111	211	211	222	221	211	111	211	211
4	111	111	111	211	221	211	211	111	222	111
5	222	221	221	211	111	211	111	111	111	211

* 222 = abundant, 221 = common, 211 = present, 111 = absent

a number of classes have been defined by using a three digit code for each character

222	Abundant	(40–100%) of rock
221	Common	(10–39%)
211	Present	(1–9%)
111	Absent	(0%)

The range of each class should be set for the particular problem in hand. Because limestones are compared by what they contain, not what they have absent in common, Jaccard's coefficient was used.

For specimens 1 and 2

$$S_J = 7/(7+8) = \tfrac{7}{15} = 0.467$$

For specimens 1 and 3

$$S_J = 8/(8+5) = \tfrac{8}{13} = 0.615$$

With this multidigit coding all digits are compared. This means there are three comparisons per petrographic constituent.

Proceeding in this fashion, we calculate all inter-sample comparisons and produce a half-matrix of similarities (table 7.13). This half-matrix can be

Table 7.13 Half-matrix of Jaccard's coefficient of association for the data from table 7.12.

Specimen					
1	—				
2	0.467	—			
3	0.615	0.400	—		
4	0.462	0.188	0.500	—	
5	0.313	0.400	0.250	0.125	—
	1	2	3	4	5

used as the starting point of a clustering procedure (see Bonham-Carter, 1967). The most similar pair of samples are linked, and then the next most similar pair and so on until all samples have been clustered in a dendrogram. Table 7.12 is a small part of a larger set of data, from which the dendrogram in figure 7.5 was produced. A clear grouping of the samples can be seen, which helped to define the limestone facies-types A–F.

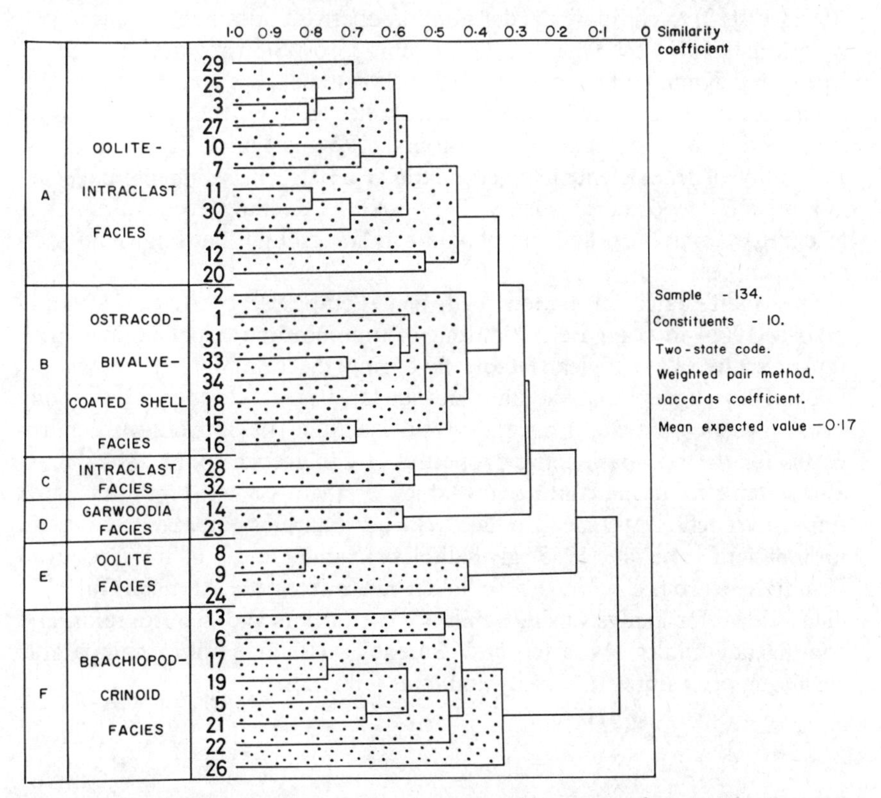

Figure 7.5 Cluster diagram, or dendrogram, of limestone facies types in terms of their constituent grain types. Provided by M. Leeder.

These similarity coefficients are very useful, but no criteria for their significance have been mentioned—because there are no rigorous ones available. However, as Bonham-Carter (1967) points out, the dendrogram is a very useful tool for compactly describing complex interrelationships between a large number of specimens.

7.4 Applications

The potential of non-parametric statistics has not been fully realised in geology, despite the excellent examples given in palaeontology by Raup and Stanley (1971), and in palaeoecology by Reyment (1971).

The χ^2-test for contingency tables is used fairly frequently in the literature. Hayami (1973) compares specimens of a Japanese scallop species with this test. Peach and Renault (1965) use contingency tables to see if there is a dependence of deposit-type on rock-type for British Columbia molybdenite occurrences.

The Wilcoxon–Mann–Whitney U-test is used by Doornkamp and King (1971, p. 41–2) to compare the distribution of areas of drainage basins in two valleys in south-west Uganda. The Kolmogorov–Smirnov test is used by Miller and Kahn (1962, appendix G) to show that there is a significant difference in the size of live *Mytilus* shells collected from rocky, muddy pool and sandy environments. Howarth and Lowenstein (1971) studied the variability of trace element determinations on stream sediments collected in a regional geochemical survey. They used the Kolmogorov–Smirnov test to compare analytical and sampling errors for ground and unground sediment samples.

Spearman's rank correlation coefficient is employed by Bates (1959) and Ghosh (1965) to compare the output of uranium and copper mines, with the presence of geological factors favouring the development of the ore body. They suggest that the mine output for a newly discovered ore body could then be predicted from their results. Batten (1973) produces dendrograms for the non-parametric properties of Cretaceous spores. His data is also suitable for use in contingency tables. Bonham-Carter (1967) calculated similarity coefficients and produced cluster diagrams for observations on foraminifera. Murray (1965) gives tables of abundances of live and dead foraminifera from a variety of sediment facies from the Arabian Gulf. His data is ideal for analysis using similarity coefficients and dendrograms, and contingency tables. Webster and Burrough (1972) employ comparable clustering procedures to classify and map soil types.

8 CONCLUSION

All the statistical tests described in the previous chapters have only involved simple arithmetic and elementary algebra. The more advanced statistical procedures require more advanced mathematics, such as matrix algebra, and would nowadays always be calculated using a computer (see Cooley and Lohnes, 1962, 1971; Overall and Klett, 1972). This text is not the place for their full mathematical treatment or description, which can be found in Miller and Kahn (1962), Krumbein and Graybill (1965), Koch and Link (1970, 1971) and Davis (1973), who devote the largest part of their books to deriving and giving worked examples of the application of more advanced statistical procedures to earth science data.

In the examples in previous chapters we have usually considered tests on one measurement, or variable, on each object (univariate statistics—for example chapters 4 and 6), or on two measurements on each object (bivariate statistics—for example chapter 5). Often we have several measurements on each object and want to consider all of them simultaneously in defining the population which the sample represents. We may define a group of fossil shells by several biometrical properties, we may compare the chemical composition of a suite of rocks or we may define drainage basins by measuring several properties. In such cases multivariate statistics are employed. There are multivariate equivalents of most of the simple statistical tests we have used. A number of procedures are also employed which are basically mathematical devices which acquire statistical overtones when used in decision making (for example, see cluster analysis and principal component analysis below).

The following paragraphs are meant to give a brief introduction to the occasions on which we might use these more advanced statistics. The short descriptions should indicate whether a technique is suitable for the problem in hand. The papers and books cited are signposts to the statistical and earth-science literature, where the full details of these procedures are described. These additional references are listed separately in a 'further reading' section at the end of the main body of references.

8.1 Partial correlation and multiple regression

The Pearson's product moment correlation coefficient (see chapter 5) expresses the relationship between a pair of variables x and y and ignores the effect on y of other variables. Often we measure a number of variables on each object and can then compute a partial correlation coefficient. This expresses the relationship between x and y with the other variables held fixed. In this way the effect of variation in the other variables, which could be masking the real relationship between x and y can be removed.

In chapter 5 a simple classical regression line was calculated for the relationship between two variables measured on each object in a sample. Multiple regression is used to explain how one variable depends on, and varies with changes in, several others. For example, we might use multiple regression to see how bulk density varies with grain size, clay content and water content of a sediment. If the dependent variable depends on two independent variables, a plane of best-fit can be calculated by the least-squares method. A similar least-squares 'surface' is calculated where there are several independent variables. A t-test or F-test can be used to assess whether the inclusion of successive independent variables in the regression significantly improves the amount of variation of the dependent variable that is explained by the multiple regression.

Draper and Smith (1966) describe the use of multiple regression methods. Krumbein (1959) gives a beautiful example of the dependence of beach firmness on four sediment properties, mean grain size, standard deviation of grain size, moisture content and porosity. His analysis showing that moisture content has the largest single effect on beach firmness, employs both partial correlation and multiple regression.

8.2 Trend surfaces

Trend-surface analysis embraces a whole series of techniques for applying a least-squares surface of best-fit to a set of locational or spatial data. The position of a value for some measurement must be defined by (x, y) co-ordinates and the value can be imagined as a topopgraphic height at that point. The trend surface defines the 'regional trend' of the measured variable in terms of x and y. For example we may wish to describe the regional variation in porosity of a sandstone body or the variation in kaolinite content of a Coal Measure tonstein.

A first-order trend surface would fit the best plane through the 'topography' of values in terms of a linear equation in x and y. A second-order surface would fit a quadratic equation and a third order surface would fit a cubic equation. Successively higher orders have been employed, but Chayes (1970) has questioned their value. The significance of a trend surface can be analysed by looking at the residuals, which are the deviation of individual values from the fitted surface.

There are numerous examples of the use of trend surfaces. Whitten (for example, 1959, 1963 and 1970) has championed their use. He has applied trend surfaces to the description of regional variations in chemical and modal properties of granite masses. Allen and Krumbein (1962) used trend surfaces to describe regional variations in the sedimentary characters of the Top Ashdown Pebble Bed of Wealden (Lower Cretaceous) age in south-east England. Harbaugh and Merriam (1968, chapter 5) review the use of trend surfaces in sedimentology, and Berry and Marble (1968) and Chorley (1972) give many geological and geographical examples.

Trend surfaces tended to be used by American and British earth scientists to the exclusion of other methods of defining regional patterns. A different method, developed by the French geological statisticians, led by Matheron, is now also available. They employ a moving-average technique called kriging, named in honour of a South African economic geologist D. G. Krige. The method of kriging is part of a new and separate group of statistical procedures referred to as 'the theory of regionalised variables'. Kriging has been used successfully in predicting gold ore reserves in South Africa and is now being applied in many other disciplines. Matheron (1971) and Watson (1971) explain the basis and use of regionalised variables, and the merits of kriging and trend surfaces are argued in a symposium on 'mathematical statistics and computer applications in ore evaluation' which was held in Johannesburg in 1966 (*Journal of the South African Institute of Mining and Metallurgy, Symposium Volume*, 1966).

8.3 Cluster analysis and discriminant analysis

We can use the related techniques of cluster and discriminant analysis to explore the similarities between objects and to define groups of objects by considering simultaneously all the variables that have been measured. For example we may want to group a set of shales in terms of their contents of a suite of trace elements or we may wish to classify a group of soil profiles in terms of a set of measured physical properties. The measured variables should be defined on the interval or ratio scale (section 7.3 briefly described clustering procedures for nominal data).

In both cluster and discriminant analysis objects are represented as points in space, defined by their values for the measured variables. Clustering proceeds by pairing nearest, which are most similar objects, to build up a group of objects until all have been grouped into clusters. If the user has a classification of his own and can define groups of objects, then discriminant analysis will give statistics to describe and distinguish these 'clouds of points'. Additional objects can be ascribed to the correct group using the calculated discriminant functions.

Purdy (1963) used a simple clustering procedure, based on correlation coefficients, to define facies types in modern Bahamian carbonate sediments.

Using the same data Parks (1966) gives a good discussion of clustering methods. Mather (1969) uses cluster analysis to group and classify third-order drainage basins from the Pennines and Anderson (1971) uses it to classify Glamorganshire soils.

Both Anderson (1958) and Li (1964) give a full discussion of discriminant statistics. Potter *et al.* (1963) use discriminant functions to separate marine and fresh-water argillaceous sediments in terms of their trace element content (particularly boron and vanadium). Davis and Sampson (1966) use this and other data in describing a computer program for discriminant analysis. Wolleben *et al.* (1968) give the data and results of a discriminant analysis applied to several Oreodont (hoofed mammal) genera.

8.4 Principal component analysis and factor analysis

We are often faced with the problem of trying simply and succinctly to describe the relationships between a complex sample of objects on which we have made many measurements. For example, how can we describe the total chemical variation in a suite of lavas, or the variation in modal composition of the sediments from a sedimentary basin? Principal component and factor analysis are designed to represent complex relationships between a large number of variables, measured on a set of objects, by simpler relationships amongst fewer variables. This reduction in the number of variables, without loss of detail about the variation, should make complex relationships more comprehensible. Various mathematical procedures are performed to describe the objects in terms of a small number of new variables. These new variables are linearly related to the original measured variables by rotation in space, and they should explain most of the sample variance in far fewer variables than were originally measured.

In principal component analysis, the new variables are usually plotted to show the relationship between samples. Factor analysis is used similarly, but is also employed to allow the relationships between the original variables to be explored. Overall and Klett (1972) describe and compare the two techniques. Harman (1967) gives details of computational procedures for factor analysis and Cattell (1965a, 1965b) gives a clear descriptive treatment. Factor analysis was initially used by psychologists and spread to other disciplines. Earth scientists were recently cautioned about its use by Matalas and Reiher (1967) and turned exclusively to principal component analysis for a time. Anderson (1958) briefly describes the statistics and Cooley and Lohnes (1962) and Walstedt and Davis (1968) give computational details of principal component analysis.

Earth scientists now regularly use both factor and principal component analysis. Imbries and Purdy (1962) employed factor analysis to classify the modern carbonate sediments found on the Great Bahama Bank. The resulting facies pattern closely resembled that obtained from Folk's limestone classification. Klovan (1966) used factor analysis to define sedimentary environments from grain size data and Spencer *et al.* (1968) used it to discover the controls on sediment chemistry and mineralogy in the Gulf of Paria.

Le Maitre (1968) used principal component analysis to describe trends in volcanic rock series. Till and Colley (1973) have suggested some possible extensions to this technique of representing petrogenetic trends. Howarth (1970) applied principal component analysis to classify samples and to define trends in Dalradian glaciomarine tills from Ireland.

8.5 Time-series analysis

Earth scientists are often analysing or describing data which varies through time. For example, we may study variation of mineralogy through a stratigraphic sequence or a soil profile, or we may record changes in spreading rates with time. In studying probability matrices and Markov chains (in section 2.2) we were looking at nominal data which varied through time. In fact, a wide variety of time-series analysis methods are used by earth scientists for all scales of measurement.

Various runs tests are used to test for time trends in a nominal or ordinal data set and are described by Miller and Kahn (1962, chapters 14 and 15) and Siegel (1956, chapter 6). Auto correlation methods are available for comparing a number of sequences of data, which have been measured on the interval or ratio scales, to see if they contain similar trends or cyclicity. This data could be trace element content of several sections or the thickness of laminae in a varied sequence. Fourier analysis is a technique that is widely used by geophysicists to analyse seismic and other records for the spectrum of frequencies they contain. It allows representation of a complex wave form by the set of simple waves that it comprises, for example see Hsu (1970). Fourier analysis can be used to analyse any cyclic sequences and also to explore the periodic functions present in a set of spatial data. The mathematics involved becomes quite complex, but Davis (1973, chapters 5 and 6) and Harbaugh and Merriam (1968, chapters 4 and 6) give very clear and understandable accounts of Fourier analysis and other types of time-trend analysis.

8.6 Postscript

Papers which are published in many of the earth-science journals employ statistical procedures, but a number of journals are worth following more closely. These are *Mathematical Geology*, *Area* and *Zeitschrift für Geomorphologie*. Most papers using statistical methods are abstracted in *Geocom Bulletin* and the appropriate sections of *Geo Abstracts*. When computer programs are needed to perform these statistical analyses they can be found in *Computer Contributions* (*Kansas Geological Survey*), *Computer Applications* (*Nottingham University*) and *Geocom Programs.*

When the reader reaches this point in the book he should have sufficient knowledge and expertise to perform statistical tests on his own data. The further reading that is referenced will lead him into more advanced fields and it is hoped that it will also lead him to start studying non-earth-science statistics texts. It is only by such study, together with lots of practice in using statistics, preferably with guidance from a colleague in a Statistics Department, that we can become competent users of statistical methods.

REFERENCES

Acton, F. S. (1966). *Analysis of Straight-Line Data*, Dover, New York.

Ager, D. V. (1973). *The Nature of the Stratigraphical Record*, Macmillan, London; Halsted Press, New York.

Ahrens, L. H. (1954). The log-normal distribution of the elements. *Geochim. cosmochim. Acta*, **6**, 121–31

Aitchison, J. and Brown, J. A. C. (1957). *The Log-normal Distribution, with Special Reference to its Use in Economics*, Cambridge University Press, Cambridge

Allen, J. R. L. (1970). Studies in fluviatile sedimentation: a comparison of fining-upwards cyclothems with special reference to coarse-member composition and interpretation. *J. sedim. Petrol.*, **40**, 298–323

Barford, N. C. (1967). *Experimental Measurements: Precision, Error and Truth*, Addison-Wesley, London

Bates, R. C. (1959). An application of statistical analysis to exploration for uranium on the Colorado plateau. *Econ. Geol.*, **54**, 449–66

Batschelet, E. (1965). *Statistical Methods for the Analysis of Problems in Animal Orientation and Certain Biological Rhythms*, American Institute of Biological Sciences, Washington

Batten, D. J. (1973). Use of palynologic assemblage types in Wealden correlation. *Palaeontology*, **16**, 1–40.

Bayly, B. (1968). *Introduction to Petrology*, Prentice-Hall, Englewood Cliffs, N.J.

Billings, G. K. and Ragland, P. C. (1968). Geochemistry and mineralogy of the recent reef and lagoonal sediments south of Belize (British Honduras). *Chem. Geol.*, **3**, 135–53

Bonham-Carter, G. F. (1967). FORTRAN IV program for Q-mode cluster analysis of nonquantitative data using IBM 7090/7094 computers. *Computer Contribution No. 17*, Kansas Geological Survey

Bozdar, L. B. and Kitchenham, B. A. (1972). Statistical appraisal of the occurrence of lead mines in the northern Pennines. *Trans. Inst. Min. Metall. Section B*, **81**. B183–88

Chayes, F. (1956). *Petrographic Modal Analysis*, Wiley, New York

Chayes, F. (1960). On correlation between variables of constant sum. *J. geophys. Res.* **65**, 4185–93

Chayes, F. (1970). Another last look at G-1/W-1. *Math. Geol.*, **2**, 207–9

Chayes, F. (1971). *Ratio Correlation: A Manual for Students of Petrology and Geochemistry*, University of Chicago Press, Chicago

Clifford, A. A. (1973). *Multivariate Error Analysis*, Applied Science Publishers, Barking

Crozier, M. J. (1973). Techniques for morphometric analysis of land slips. *Z. Geomorph. N.F.*, **17**, 78–101

Curl, R. C. (1966). Caves as a measure of karst. *J. Geol.*, **74**, 798–830

Curtis, C. D. (1969). Trace element distribution in some British Carboniferous sediments. *Geochim. cosmochim. Acta*, **33**, 519–23

Davis, J. C. (1973). *Statistics and Data Analysis in Geology*, Wiley, New York

Dixon, W. J. and Massey, F. J. (1957). *Introduction to Statistical Analysis*, McGraw-Hill, New York

Doornkamp, J. C. and King, C. A. M. (1971). *Numerical Analysis in Geomorphology: An Introduction.* Arnold, London

Draper, N. R. and Smith, H. (1966). *Applied Regression Analysis*, Wiley, New York

Fairbairn, H. W. (1951). A co-operative investigation of precision and accuracy in chemical, spectrochemical and modal analysis of silicate rocks. *Bull. U.S. geol. Surv.*, **980**, 1–24

Feller, W. (1968). *An Introduction to Probability Theory and its Applications. Volume 1*, Wiley, New York

Fisher, R. A. (1956). *Statistical Methods and Scientific Inferences*, Oliver & Boyd, Edinburgh

Fisher, R. A. and Yates, F. (1948). *Statistical Tables for Biological, Agricultural and Medical Research*, Oliver & Boyd, Edinburgh

Flanagan, F. J. (1973). 1972 values for international geochemical reference samples. *Geochim. cosmochim. Acta*, **37**, 1189–200

Fleischer, M. (1965). Summary of new data on rock samples G-1 and W-1, 1962–1965. *Geochim. cosmochim. Acta*, **29**, 1263–83

Fleischer, M. and Stevens, R. E. (1962). Summary of new data on rock samples G-1 and W-1. *Geochim. cosmochim. Acta*, **26**, 525–43

Fryer, H. C. (1966). *Concepts and Methods of Experimental Statistics*, Allyn & Bacon, Rockleigh, N.J.

Ghosh, A. K. (1965). A statistical approach to the exploration of copper in the Singhbhum shear zone, Bihar, India. *Econ. Geol.*, **60**, 1422–30

Gray, A. and Matthews, G. B. (1922). *A Treatise on Bessel Functions and Their Applications to Physics*, Macmillan, London

Griffiths, J. C. (1967). *Scientific Method in Analysis of Sediments*, McGraw-Hill, New York

Gunatilaka, H. A. and Till, R. (1971). A precise and accurate method for the quantitative determination of carbonate minerals by X-ray diffraction using a spiking technique. *Min. Mag.*, **38**, 481–7

Harbaugh, J. W. and Bonham-Carter, G. (1970). *Computer Simulation in Geology*, Wiley, New York

Harbaugh, J. W. and Merriam, D. F. (1968). *Computer Applications in Stratigraphic Analysis*, Wiley, New York

Hayami, I. (1973). Discontinuous variation in an evolutionary species, *Cryptopecten vesiculosus*, from Japan. *J. Palaeont.*, **47**, 401–20

Hellewell, E. G. and Myers, J. O. (1973). Measurement and analysis of *in situ* rock densities in British Carboniferous and Permo-Triassic. *Trans. Inst. Min. Metall. Section B*, **82**, B51–60

Howarth, R. J. and Lowenstein, P. L. (1971). Sampling variability of stream sediments in broad-scale regional geochemical reconnaissance. *Trans. Inst. Min. Metall. Section B*, **80**, B363–72

Kahn, J. S. (1956). The analysis and distribution of the properties of packing in sand size sediments. 2. The distribution of the packing measurements and an example of packing analysis. *J. Geol.*, **64**, 578–606

Kendall, M. G. (1970). *Rank Correlation Methods*, Griffin, London

Kermack, K. A. and Haldane, J. B. S. (1950). Organic correlation and allometry. *Biometrika*, **37**, 30–41

Koch, G. S. and Link, R. F. (1963). Distribution of metals in the Don Tomas vein, Frisco Mine, Chihuahua, Mexico. *Econ. Geol.*, **58**, 1061–70

Koch, G. S. and Link, R. F. (1970). *Statistical Analysis of Geological Data*, Wiley, New York

Koch, G. S. and Link, R. F. (1971). *Statistical Analysis of Geological Data. Volume II*, Wiley, New York

Krinsley, D. (1960). Magnesium, strontium and aragonite in the shells of certain littoral gastropods. *J. Palaeont.*, **34**, 744–55

Krumbein, W. C. (1967). FORTRAN IV computer programs for Markov chain experiments in geology. *Computer Contribution No. 13*, Kansas Geological Survey

Krumbein, W. C. and Graybill, F. A. (1965). *An Introduction to Statistical Models in Geology*, Wiley, New York

Krumbein, W. C. and Pettijohn, F. J. (1938). *Manual of Sedimentary Petrography*, Appleton-Century Crofts, New York

Lees, A., Buller, A. T. and Scott, J. (1969). Marine carbonate sedimentation processes, Connemara, Eire. An interim report. *Reading University Geological Report No. 2*

Li, J. C. R. (1964). *Statistical inference, I*, Edwards Brothers, Ann Arbor

MacDonald, R. and Bailey, D. K. (1973). The chemistry of the peralkaline oversaturated obsidians. *Prof. Pap. U.S. geol. Surv.*, **440–N–1**

Mandel, J. (1964). *The Statistical Analysis of Experimental Data*, Wiley, New York

Mardia, K. V. (1972). *Statistics of Directional Data*, Academic Press, London

Miall, A. D. (1973). Markov chain analysis applied to an ancient alluvial plain succession. *Sedimentology*, **20**, 347–64

Miller, R. L. and Kahn, J. S. (1962). *Statistical Analysis in the Geological Sciences*, Wiley, New York

Moroney, M. J. (1970). *Facts from Figures*, Penguin, Harmondsworth

Murdoch, J. and Barnes, J. A. (1974). *Statistical Tables for Science, Engineering, Management and Business Studies*, 2nd revised edition, Macmillan, London

Murray, J. W. (1965). The foraminiferida of the Persian Gulf. 2. The Abu Dhabi Region, Palaeogeog, *Palaeoclim. Palaeoecol.*, **1**. 307–32

Parkinson, D. (1954), Quantitative studies of brachipods from the lower Carboniferous reef limestones of England. I. *Schizophonia resupinata* (Martin). *J. Palaeont.* **28**, 367–81

Peach, P. A. and Renault, J. R. (1965). Statistical analysis of some characteristics of British Columbia molybdenite occurrences. *Econ. Geol.*, **60**, 1510–15

Pearson, E. S. and Hartley, H. O. (1969). *Biometrika Tables for Statisticians*, Cambridge University Press, Cambridge

Raup, D. M. and Stanley, S. M. (1971). *Principles of Palaeontology*, Freeman, San Francisco

Read, W. A. (1969). Analysis and simulation of Namurian sediments in Central Scotland using a Markov-process model. *Math. Geol.*, **1**, 199–219

Read, W. A. and Dean, J. M. (1967). A quantitative study of a sequence of coal-bearing cycles in the Namurian of central Scotland, 1. *Sedimentology*, **9**, 137–56

Read, W. A. and Dean, J. M. (1968). A quantitative study of a sequence of coal-bearing cycles in the Namurian of central Scotland, 2. *Sedimentology*, **10**, 121–36

Reyment, R. A. (1966). Preliminary observations on gastropod predation in the western Niger delta. *Palaeogeog. Palaeoclim. Palaeoecol.*, **2**, 81–102

Reyment, R. A. (1971). *Introduction to Quantitative Palaeoecology*, Elsevier, Amsterdam

Royal Society (1971). *Quantities, Units and Symbols*, Royal Society, London

Selley, R. C. (1969). Studies of sequence in sediments using a simple mathematical device. *Q. Jl geol. Soc. Lond.*, **125**, 557–81

Siegel, S. (1956). *Non-parametric Statistics for the Behavioural Sciences*, McGraw-Hill, New York

Sneath, P. H. A. and Sokal, R. R. (1973). *Numerical Taxonomy: The Principles and Practice of Numerical Classification*, Freeman, San Francisco

Sokal, R. R. and Sneath, P. H. A. (1963). *Principles of Numerical Taxonomy*, Freeman, San Francisco

Spiegel, M. R. (1961). *Schaum's Outline of Theory and Problems of Statistics*, McGraw-Hill, New York

Stevens, R. E. and Niles, W. M. (1960). Second report on a co-operative investigation of the composition of two silicate rocks. *Bull. U.S. geol. Surv.*, **1113**, 3–43

Taylor, R. K. (1973). Compositional and geotechnical characteristics of a 100-year old colliery spoil heap. *Trans. Inst. Min. Metall. Section A*, **82**, A1–14

Till, R. (1971). Are there geochemical criteria for differentiating reef and non-reef carbonates? *Bull. Am. Ass. Petrol. Geol.*, **55**, 523–8

Till, R. (1973). The use of linear regression in geomorphology. *Area* 5, 303–8

Till, R. and Colley, H. (1973). Thoughts on the use of principal component analysis in petrogenetic problems. *Math. Geol.*, **5**, 341–50

Till, R., Hopkins, D. T. and McCann, C. (1972). A collection of computer programs in BASIC for use in geology and geophysics. *Reading University Geological Report No.* 5

Vistelius, A. B. (1970). Statistical models of silicate analysis and results of investigation of G-1 and W-1 samples. *Math. Geol.*, **2**, 1–14

Vistelius, A. B., Ivanov, D. N., Kuroda, Y. and Ruiz Fuller, C. (1970). Variations of modal composition of granitic rocks in some regions around the Pacific. *Math. Geol.*, **2**, 63–80

Von Mises, R. (1939). *Probability, Statistics and Truth*, Hodge, London

Von Mises, R. (1964). *Mathematical Theory of Probability and Statistics*, Academic Press, New York

Wadati, K. (1967). Earthquake, Depth of (Deep-focus earthquake), in *International Dictionary of Geophysics* (ed. S. K. Runcorn), Pergamon, London p. 389–92

Watson, G. S. (1966). The statistics of orientation data. *J. Geol.*, **74**, 786–97

Webster, R. and Burrough, P. A. (1972). Computer based soil mapping of small areas from sample data. *J. Soil Sci.*, **23**, 210–34

Yevjevich, V. (1971). *Probability and Statistics in Hydrology*, Water Resources Publications, Colorado

Yule, G. U. and Kendall, M. G. (1953). *An Introduction to the Theory of Statistics*, Griffin, London

FURTHER READING

Allen, P. and Krumbein, W. C. (1962). Secondary trend components in the Top Ashdown Pebble Bed: A case history. *J. Geol.*, **70**, 507–38

Anderson, A. J. B. (1971). Numerical examination of multivariate soil samples. *Math. Geol.*, 3, 1–14

Anderson, T. N. (1968). *An Introduction to Multivariate Statistical Analysis*, Wiley, New York

Berry, B. J. L. and Marble, D. F. (1968). *Spatial Analysis, A Reader in Statistical Geography*, Prentice-Hill, Englewood Cliffs, N.J.

Cattell, R. B. (1965a). Factor analysis: An introduction to essentials: I. *Biometrics*, **21**, 190–215

Cattell, R. B. (1965*b*). Factor analysis: An introduction to essentials: II. *Biometrics*, **21**, 405–35

Chayes, F. (1970). On deciding whether trend surfaces of progressively higher order are meaningful. *Geol. Soc. Am. Bull.*, **81**, 1273–78

Chorley, R. J. (1972). *Spatial Analysis in Geomorphology*, Methuen, London

Cooley, W. W. and Lohnes, P. R. (1962). *Multivariate Procedures for the Behavioural Sciences*, Wiley, New York

Cooley, W. W. and Lohnes, P. R. (1971). *Multivariate Data Analysis*, Wiley, New York

Davis, J. C. (1973). *Statistics and Data Analysis in Geology*, Wiley, New York

Davis, J. C. and Sampson, R. J. (1966). FORTRAN II program for multivariate discriminant analysis using an IBM 1620 computer. *Computer Contribution No.* 4, Kansas Geological Survey

Draper, N. R. and Smith, H. (1966). *Applied Regression Analysis*, Wiley, New York

Harbaugh, J. W. and Merriam, D. F. (1968). *Computer Applications in Stratigraphic Analysis*, Wiley, New York

Harman, H. H. (1967). *Modern Factor Analysis*, University of Chicago Press, Chicago

Howarth, R. J. (1970). Principal component analysis of the geochemistry and mineralogy of the Portaskaig tillite and Kiltyfanned Schist (Dalradian) of Co. Donegal, Eire, *Math. Geol.*, **2**, 285–302

Hsu, H. P. (1970). *Fourier Analysis*, Simon & Schuster, New York

Imbries, J. and Purdy, E. G. (1963). Classification of modern Bahamian carbonate sediments, in *Classification of Carbonate Rocks, a Symposium* (ed. W. E. Ham), American Association of Petroleum Geologists, Memoir 1, 253–72

Klovan, J. E. (1966). The use of factor analysis in determining depositional environments from grain size distributions. *J. Sedim. Petrol.*, **36**, 115–25

Koch, G. S. and Link, R. F. (1970). *Statistical Analysis of Geological Data*, Wiley, New York

Koch, G. S. and Link, R. F. (1971). *Statistical Analysis of Geological Data. Volume II*, Wiley, New York

Krumbein, W. C. (1959). The 'sorting out' of geological variables illustrated by regression analysis of factors controlling beach firmness. *J. Sedim. Petrol.*, **29**, 575–87

Krumbein, W. C. and Graybill, F. A. (1965). *An Introduction to Statistical Models in Geology*, Wiley, New York

Le Maitre, R. W. (1968). Chemical variation within and between volcanic rock series—a statistical approach. *J. Petrol.*, **9**, 220–52

Li, J. C. R. (1964). *Statistical inference, II*, Edwards Brothers, Ann Arbor

Matalas, N. C. and Reiher, B. J. (1967). Some comments on the use of factor analysis. *Wat. Resour. Res.*, **3**, 213–23

Mather, P. M. (1969). Cluster analysis. *Computer Applications No.* 1, Nottingham

Matheron, G. (1971). The theory of regionalised variables and its applications. *Cah. Cent. Morph. Math.* No. 5

Miller, R. L. and Kahn, J. S. (1962). *Statistical Analysis in the Geological Sciences*, Wiley, New York

Overall, J. E. and Klett, C. J. (1972). *Applied Multivariate Analysis*, McGraw-Hill, New York

Parks, J. M. (1966). Cluster analysis applied to multivariate geological problems. *J. Geol.*, **74**, 703–15

Potter, P. E., Shimp, N. F. and Witters, J. (1963). Trace elements in marine and fresh-water argillaceous sediments. *Geochim. cosmochim. Acta*, **27**, 669–94

Purdy, E. G. (1963). Recent calcium carbonate facies of the Great Bahama Bank. 1. Petrography and reaction groups. *J. Geol.*, **71**, 415–25

Siegel, S. (1956). *Non-parametric Statistics for the Behavioural Sciences*. McGraw-Hill, New York

Spencer, D. W., Degens, E. T. and Kulbicki, G. (1968). Factors affecting element distributions in sediments, in *Origin and Distribution of the Elements* (ed. L. H. Ahrens), Pergamon, London, 981–98

Till, R. and Colley, H. (1973). Thoughts on the use of principal component analysis in petrogenetic problems. *Math. Geol.*, **5** 341–50

Wahlstedt, W. C. and Davis, J. C. (1968). FORTRAN IV program for computation and display of principal components. *Computer Contribution No. 21*, Kansas Geological Survey

Watson, G. S. (1971). Trend-surface analysis. *Math. Geol.*, **3**, 215–26

Whitten, E. H. T. (1959). Compositional trends in a granite: Modal variation and ghost stratigraphy in part of the Donegal granite, Eire. *J. geophys. Res.*, **64**, 835–48

Whitten, E. H. T. (1963). Application of quantitative methods in the geochemical study of granite massifs. *R. Soc. Can. Spec. Publs No. 6*, 76–123

Whitten, E. H. T. (1970). Orthogonal polynomial trend surfaces for irregularly spaced data. *Math. Geol.*, **2**, 141–52

Wollenben, J. A., Pauken, R. J. and Dearien, J. A. (1968). FORTRAN IV program for multivariate palaeontologic analysis using an IBM system/360 model 40 computer. *Computer Contribution No. 20*, Kansas Geological Survey, 1–12

INDEX

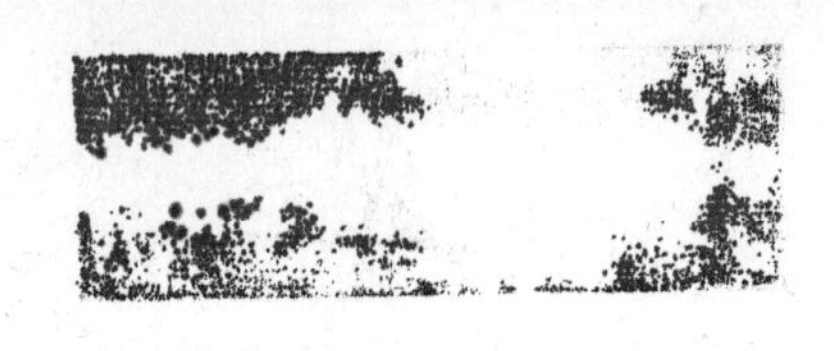